BEING BIOLOGICAL
Human Meaning
in the Age of Neuroscience

Michael A. Fuller

2022

© 2022 by Michael A. Fuller

Printed in the United States

Fuller, Michael A.

Being Biological: Human Meaning in the Age of Neuroscience

All rights reserved. No part of this book may be reproduced, adapted, stored in a retrieval system or transmitted by any means, electronic, mechanical, photocopying or otherwise without the prior written approval of the author.

BISAC: SCI089000
        LIT006000

Includes bibliographical references and index

ISBN 979-8-9873317-0-5 (Hard cover);
       979-8-9873317-1-2 (Paperback);
       979-8-9873317-2-9 (eBook)

Index by the author

To Kathy

Table of Contents

	List of Figures	iii
	Introduction	1
1	REFRAMING	12
	Neurons and Neural Networks	
	The Predictive Brain	
	From Perceptrons to a Human-built World	
2	CONNECTIONS: INTRODUCING THE BRAIN	49
	Introducing the Brain	
	The Brainstem	
	The Forebrain	
	The Cerebral Cortex	
	Networks in the Cortex	
3	SEEING THE LIGHT	68
	The Eye and the Retina	
	The Lateral Geniculate Nucleus and the Primary Visual Cortex	
	From the Primary Visual Cortex to the Visual System	
	Retrospective and Prospective	
4	AFFECT: THE FEEL OF BEING	96
	A Brief Digression on Not Including a Survey of the History of Emotion	
	The Three Tiers of Affective Organization	
5	DEVELOPMENT: BUILDING A BRAIN, SHAPING A SELF	135
	In the Beginning: The Early Development of the Human Brain	
	Getting it Together: Core Functional Networks	
	The "Self" in the System	
6	BUILDING A WORLD OF MEANING: THE NEURAL DYNAMICS OF MEMORY	169
	The Basic Model	
	Such Stuff as Dreams are Made on	

      Memory and the Medial Temporal Lobe
      Sleep and Memory Consolidation
      Semantic Memory and the Construction of Meaning

7 BEING BIOLOGICAL   218
      Phenomenological / Connections
      A Poetry of Embodiment

PERMISSIONS   241

BIBLIOGRAPHY   249

INDEX   269

List of Figures

Figure 1.1:	The Neuron
Figure 1.2:	A Grid of Neurons to Recognize Letters
Figure 1.3a:	"Winner takes all" Neural Network
Figure 1.3b:	"Winner takes all" Neural Network Trial Results
Figure 1.4:	Model of Neural Network connecting the Lateral Geniculate
Figure 1.5:	Kohonen Self-organizing Map Network
Figure 1.6:	"Mexican Hat" Weight function for Lateral Connections in SOM
Figure 1.7a:	Architecture of the LISSOM model
Figure 1.7b:	LISSOM Topological Feature Map Model for Line Orientation in Hypercolumns in V1
Figure 1.8:	Model of Recurrent Connectivity in the Visual Cortices
Figure 1.9:	Attractor Basin
Figure 1.10:	Figure-ground Discrimination
Figure 1.11:	Attractor Basin Landscape
Figure 1.12:	The Adaptive Resonance Model
Figure 2.1:	The Human Brain
Figure 2.2:	Medial View of the Brain
Figure 2.3:	The Lateral View of the Neocortex
Figure 2.4:	Names of Directions for the Brain
Figure 2.5:	The 6-Layered Neocortex
Figure 2.6:	Functional Roles Identified for Brodmann Areas
Figure 2.7a:	Sensory Homunculus
Figure 2.7b:	Motor Homunculus
Figure 2.8:	White Matter Tracts
Figure 2.9:	Developmental Changes in Major Brain Networks
Figure 2.10:	Major Networks in the Brain
Figure 3.1:	The Human Eye
Figure 3.2:	The Neuronal Structure of the Retina
Figure 3.3:	The Fovea
Figure 3.4:	Using Eye-tracking to Record Saccades
Figure 3.5:	Photoreceptor Response Sensitivity

Figure 3.6:	The Center-Surround Model
Figure 3.7:	Image of Filtering to enhance edges in Ganglion Cells
Figure 3.8:	Binocular Visual Pathways
Figure 3.9:	Orientation Selectivity in V1
Figure 3.10:	Ocular Dominance Columns
Figure 3.11:	Increasing Complexity of Feature-Map Structural Units
Figure 3.12:	The Visual Cortices
Figure 3.13:	Reciprocal Connectivity to the Visual Cortex
Figure 3.14:	Neuronal Control Networks and Hubs
Figure 3.15:	The Dorsal and Ventral Streams
Figure 4.1:	Affective Hierarchies
Figure 4.2:	The Human Parenting Network
Figure 4.3a:	The Visual Perceptual Hierarchy
Figure 4.3b:	The Body/Emotion Hierarchy
Figure 4.4:	Affective Networks of the Brain
Figure 5.1:	The Fetal Forming of the Layers of the Cortex
Figure 5.2:	Developmental Timeline
Figure 5.3:	Neuronal Developmental Timeline
Figure 6.1	An Artificially Generated Dendrogram
Figure 6.2:	Catastrophic Interference in Learning
Figure 6.3:	Distribution of Memory in the Brain
Figure 6.4:	The Medial Temporal Lobe Structure
Figure 6.5:	The Two Pathways in the Hippocampus
Figure 6.6:	The Mossy Fiber- CA3 Synapse
Figure 6.7:	The Connections to CA3 in the Hippocampus
Figure 6.8:	The Sleep Cycle
Figure 6.9:	Memory Consolidation
Figure 6.10	Neural Networks to Support Language Perception and Production
Figure 6.11:	Dependency Parsing
Figure 6.12:	Phrase-Structure Parsing
Figure 6.13:	Connectivity in the Language System

Introduction

Letting Go of the Self

In the twilight hours of the night that Siddhartha Gautama attained nirvana, he saw all the causes that drove his hopes, desires, and fears, all the external and internal impulses that shaped who he had become. Deeper into the night, he saw the vast net of causes and conditions in which humans had been caught throughout the centuries, the ceaseless churning of the material world that had led to exactly how things were at precisely that moment. He realized then that he had no self, no core of being that stood outside those self-generating cycles of cause and effect. He attained nirvana—the extinction of the self and of those cycles of causation—and became what he now knew he always had been, the Buddha.

This story presents a paradox of identity. In the Buddhist world of inescapable causation, the self dissolves into the universal network of becoming, only to return, exactly the same yet entirely different because the very nature of identity and the system in which it is embedded have shaken off old ontological expectations. "First there is a mountain, then there is no mountain, then there is."[1] As the science of the brain steadily encroaches on our sense of freely formed

---

[1] This is the singer Donovan's famous rephrasing of a Buddhist account of the experience of categories sliding away and returning: The Chan master Qingyuan Weixin 青原惟信 (9th century) observed, "Thirty years ago, before I started Chan meditation, I saw mountains as mountains and rivers as rivers. Later, after I had some experience [with awakening] and had entered somewhat, I saw mountains but they weren't mountains and the rivers were not rivers. Now I have found some ease [in the Way] and as before, I see mountains as just mountains and rivers as just rivers." (Puji, ed., *Wudeng huiyuan* 五燈會元, collated by Su Yuanlei [Beijing: Zhonghua, 1984] 3.17.1135.) D. T. Suzuki included this saying in his *Essays in Zen Buddhism* (New York: Grove Press, 1949), p. 24.

identities, we as a culture will be facing such a night of dissolution of a subjective self still grounded in the remnants of old idealist dreams. I seek, at this juncture, to offer a vision of identity transformed, renewed, and deepened through dispersion into the vast, rich network of biological patterns out of which the self as we know it arises. I make no claims about the status of the soul: I simply do not know; my project is instead to account for what we can know of human experience based on what we can know empirically of the processes of the material world. Central to this empirical knowledge of the relationship of human experience to the phenomenal realm (the manifold of experience) is what neuroscience can tell us about perception, cognition, emotion, and memory.[2]

I write this book for readers who share a commitment to finding meaning within the nuanced patterns of the world of human experience. I hope to show that neuroscientific accounts are consistent with the complexity of this human world. That mountain, in the end, is still that mountain. The end of "King Lear" remains a crushing tragedy; Bach's "Mass in B-Minor" continues to touch the ineffable: the hopes and sorrows of subjective experience are real.[3] What is "in our head," being in our head, is real: what else would it be? Our subjective life at times may seem an errant fantasy or will-o-wisp of desire or fear, but it arises from processes of imagination, thought and feeling that unfold in the brain. I suggest that engagement with neuroscientific models offers us a different understanding of our inner life and all that we label the subjective self. The biologically conceived self is the totality of neural structures—the channels and connections constructed over a lifetime of experience—and is as complex,

---

[2] While I write this book to present an approach to contemporary neuroscience that highlights aspects that are directly relevant to humanists, there are in fact excellent general introductions to neuroscience. One that I have found particularly clear and judicious in its details and assessments is Bernard J. Baars and Nicole M Gage, *Fundamentals of Cognitive Neuroscience: A Beginner's Guide* (Boston: Academic Press, 2013).

[3] To be more precise at the outset: throughout the book, my claims about the "real" are limited. I accept Immanuel Kant's skepticism about what we can know: he famously argued that we can have no direct access to things in themselves—as they really are—nor can we truly know the self that orders the world of experience within which we live. All we can know is this "manifold of experience," the totality of objects and events that present themselves to us. I accept this constraint, and it is this circumscribed, *phenomenal* realm to which I refer.

and also as fragile and conflicted, as the self explored in the cultural traditions of the world over the past four thousand years.

Shifting Perspectives

The project of thinking of the self in biological terms is not easy, in part because biology is not just a matter of the test tubes and microscopes that, in the popular imagination, are the mainstay of research. Contemporary biology draws extensively on abstract elements of physics, chemistry, statistics, and—for neural networks in particular—on a branch of mathematics called complex systems theory. There are few simple explanations for biological phenomena in this hybrid discipline because, in fact, Nature is not simple.

The complexity and diversity of biology as a discipline reflect the complexity of its object of study. Its sub-disciplines focus on the myriad different levels of organization out of which organisms arise and within which they live. The diverse components and dynamics of life at the molecular level produce cellular structures with distinctive emergent properties deriving from their molecular organization. Cells form into tissues, and tissues form organs, and on and on until we have individual members of a species. The individuals then participate in larger social organizations and ecosystems with emergent properties deriving from the collective behavior of those individuals.[4] The success of these emergent group properties in turn biases the genetic survival of those systemic features that, at the lower levels of organization, lead to successful behavior. Biologists study all of this with a corresponding range of research procedures and modes of thought that apply most effectively at the distinctive levels of biological organization. There is no one Biology anymore.

Exploring the mind and self as (ultimately) biological systems requires an awareness of—and respect for—the power of the distinctive paradigms designed to articulate the organizational and behavioral logic of progressively higher levels of structuring. Neuroscience from the beginning has been in active conversation with cognitive science, neurology, and cognitive linguistics. More

---

[4] An emergent property is precisely one that appears through the interaction of elements in a system and is not the property of any individual element.

recently, as neuroscience increasingly explores ever higher levels of brain organization, it has begun to engage with more traditional fields within the broad discipline of psychology and to test its findings and models against those of these long-established fields of study. Still, neuroscience—even in conversation with other fields—is not sufficient in itself. The disciplines needed to articulate adequately the organizational logic of a biologically conceived self extend far beyond biology. Just as chemistry is built upon physics but has its own models, procedures, and objects of inquiry distinct from physics, so, too, exploring large-scale patterns of human experience—even with a commitment to its final biological grounding—must draw upon conceptual tools and models that are different from those of neuroscience.

How, then, do the humanities participate in this cross-disciplinary conversation? Here I think it is crucial for us in the humanities to see our own work, our own domains of inquiry—literature, history, and philosophy, etc.— and our own theoretical models as part of this project of consilience that connects layer upon layer of disciplinary accounts that build upon biological models of brain organization. If we take up the challenge of seeing ourselves as part of a larger intellectual enterprise of explaining what we can of human experience within the dynamics of the material world and think of our own self-understanding in relation to other models of human meaning that arise out of the neuroscientific project, we also challenge those other models to acknowledge and accommodate the complexities of experience explored in the humanities. Artists and scholars in the humanities present a domain of experience that must be part of the human story in its own terms as well as understood as relying on a substructure of biological foundations. In this newly emerging engagement, the accounts that humanists have developed can only be enriched and deepened by their connections to the substratum of biological processes, while at the same time, making distinctive, signal contributions to the larger project.

The Argument of the Book

The argument of the book—to put it in one long sentence—is that the particular developmental logic that gives great weight to affective mechanisms in the shaping of a memory system, in which the objects and events in memory are not discrete but defined through mutual differentiation, produces an inner world of meaning bound to the body and to experience that can serve as the substructure for discourses of meaning and selfhood in the humanities. I begin with the structural logic of representation in the brain, in which contemporary neuroscientific paradigms mirror central theoretical positions in the humanities. That is, the brain models the world and the human body through a hierarchy of neural networks that define increasingly higher dimensional representational spaces. Given the mathematics of neural networks, the brain identifies the "objects" of experience not individually and atomically but as parts of a system of mutually differentiated activation states.[5] Thus, the brain uses a structuralist model of "meaning by difference." However, the human brain was designed for collective survival, not truth, even if having an accurate model of expectations about the body and the world is very useful for survival. The particular needs and limitations of the body as well as roles of the objects of the environment in contributing to survival crucially shape the structure of the representational systems of the brain. Hence, one key argument of this book at the level of the internal organization of neural networks is that these networks are *poststructuralist*: they are not arbitrary in the distinctions they make but instead are driven by need and assessments of power and possibility.

A second key argument of the book that derives from the first is that the simple divide between subjective emotion and objective cognition is not just wrong but harmful. The representational structures of the brain are not neutral. Layers of neural networks, beginning with the sensory and somatosensory cortices, extract categories of regularities in the environment and the body that a complex system of neuromodulators produced by subcortical regions define as significant. The organization of the higher order structures is determined by these subcortical systems interacting with early experience of the world and the body. What we call emotions, based on an individual's experiential articulation of the

---

[5] Of course in the mature brain, "object memory" is sufficiently highly articulated that individual objects can be introduced into the structure with most of their specificity intact so that they appear for all practical purposes as atomic, self-subsisting things that seemingly do not draw on their places in the structure to define them.

subcortical systems, are as objective as any higher order biological system. They are real.

The third key argument considers a developmental component of the creation of the self. We are born with a particular body to particular parents in a particular place with its given culture, society, and built environment, all of which impinge as patterns upon the brain as it seeks to extract the internal and external regularities out of which it will build the self. This "self" is the emergent totality of systems that shape response and action.[6] However, the brain at birth is not mature. Instead, the cortex develops from the back of the brain (the sensory cortices) forward. At birth, the sensory cortices are "experience-ready:" they are structured to be ready to rely on information from the already-mature primary process emotions to build neural networks that are "experience-dependent." Even the establishment of the long-range connections of the sensory cortices to higher cortical processing areas is experience-dependent. The early self that arises out of this initial affect-laden mapping of the world demonstrates remarkable persistence in shaping the self throughout life.

A fourth key argument of the book derives directly from the first three: memory and language (as one component of memory) are not subject to the infinite regress of the referent but deeply embodied. Beginning in early infancy, rapidly developing cortical and subcortical networks extract meaningful patterns re-presenting emergent objects and events encountered in experience. Additional networks bind these patterns to sounds, which eventually become words, and other networks construct the syntactic features of the infant's language. Neuroscientific research into the development of memory and language has made significant progress in outlining the ontogeny of "meaning" in general and of language in particular and of their connections to other aspects of infant development. The adult versions, needless to say, grow out of the initial infant structures and processes and provide us with a supple theory of reference for language that

---

[6] Here I confess I find the basic psychological categories of the classical Chinese tradition helpful in providing a model formally similar to that of neuroscience. The self is the *xing* 性, or "nature;" it has a structure of commitments concerning the world, the *zhi* 志, or "resolve," and individuated responses that can be considered either *qing* 情, "feelings" or *yi* 意, "intentions." See Chapter 7 for a detailed discussion of these terms.

remains entirely within the material realm (and thus makes no claims beyond its limited empirical applicability). There is little cost to humanists in accepting such a neuroscientific theory of reference. Quite to the contrary, there is much to be gained. Returning language to the body and to history (and bracketing metaphysical concerns for the Absolute) turns the focus of meaning to the particularities of experience.

Taken together, the points I underscore present a model of substantial, this-worldly meaning in which the self—through its biology and its history—participates in a larger world of the patterns that permeate phenomenal experience. I argue moreover that this model of meaning as a complex web of relationships is implicit in a synthesis of the current state of the fields of computational, affective, and developmental neuroscience along with the neuroscience of memory. The humanities can—and should—draw on this emerging synthesis to participate in exploring the nature of the construction of meaning in conversation with neuroscience.

The Structure of the Book

This book presents in stages a model of meaning based on my selective synthesis of neuroscientific perspectives. The first step is large because it requires an imaginative reenvisioning of our encounter with the world in all its aspects: it asks us to turn away from our sense of the givenness of experience. We need to take seriously the specifically biological reconstruction of the encountered world and not just as a metaphor. In this account, nothing that we know, hope, feel, or remember is what it seems, as it fragments into activations of networks of neurons spread across the brain. The first chapter begins this shift in the conceptual framework by introducing the basic elements of the biological substructure of experience. The chapter first introduces neurons themselves, moves to simple mathematical models of neurons and then explores the modeling of ever more complex networks of neurons. These models for the behavior of neural networks are central to contemporary neuroscientific theorizing and provide the foundation on which the rest of the book is built. However, recent theorizing about the computational logic of the brain has come to stress that brains are built not simply to react to input but to *predict* what is most likely the cause of the present input and to then plan responses to that cause in a timely and accurate manner. Therefore, I next introduce the increasingly important perspective of "predictive processing" in which the goal of the neural organization of the neocortex is to build a model of the patterns of the world so that, once the model is developed, most of our understanding of the world in our seemingly immediate encounter with it is in fact a form of recalling past patterns, developing expectations and plans based on predicting the present, and checking for errors in those predictions.

The second chapter shifts from the tightly structured world of neural networks to an overview of the organization of the brain and its component systems within which neuronal networks take shape and connect with one another. The first goal of this chapter is simply to introduce the names of the regions of the brain that appear throughout the rest of the book so that the reader can know where the regions are, what systems they are a part of, and what their usually-described functions are. The second goal is to stress that most brain functions—from emotion to vision to planning—are an ensemble performance in which areas and nuclei from all over the brain participate. Toward this end, the last part of the chapter introduces models of connectivity in the brain that have been developing in research on network neuroscience, the study of the ways in which the discrete local neural networks interconnect to form large-scale, long-distance networks whose formal structures are studied in graph theory.[7]

The third chapter draws on the modeling from chapter one to present an overview of the visual system, the best-studied of all the sensory systems in the brain. The chapter takes another step in the fragmentation of experience. The details of the processing of visual information starting in the retina dispel the illusion of the immediacy and simple coherence of what one sees when one opens one's eyes. Starting with the types of cells in the retina itself, the chapter makes clear that the evolved specificity of biological systems matters, but also that human systems, while distinct in many ways, are part of an evolutionary continuum. The chapter traces the ways in which visual data, already transformed in

---

[7] Graph theory is the branch of mathematics that provides the analytic models for social network analysis (SNA), with which humanists are becoming increasingly familiar.

the retina, undergo layer upon layer of extractions of ever higher order patterns to identify significant objects and events. However, the end of the chapter returns to the theme of predictive processing and presents a detailed view of vision as half-seen, half predicted: the predictive model is both computationally efficient and—crucially—accounts for the massive feedback connections at every level within hierarchies of neural networks. Higher levels of the visual system develop expectations about the world (or at least about the patterns of activations they have received) and draw on these expectations to interpret and—through feedback loops—to shape the data that comes to them. The chapter presents a concrete example of how we see and live in the "remembered present."

Chapter Four, on emotion, continues the exploration of the ways in which the brain shapes the neocortical modeling of the world it encounters. The brain at birth knows almost nothing about the body or the world. Yet the processes of response developed through the hierarchies of neural networks in the neocortex require specifically neocortical models of the body and the world in order to anticipate, assess, and respond. The affective system, built upon bodily connections to the structures in the brain stem, is responsible for providing information on bodily needs to the experience-dependent cortical networks both for immediate action and as a crucial way to determine what is important in the patterns of encounter. A central question for affective neuroscience is how the cortical mappings of the midbrain systems produce the emergent properties we observe in what we know as emotions. That is, humans are social creatures, yet sociality is not directly in the genes but is the tertiary-level emergent result of early system dynamics. How do the cortical responses to the midbrain systems lead to this result? The chapter sets out the three levels of the affective system and traces their interaction as well as their impact on other cortical networks for perception, cognition, planning, and memory.

While the chapters on vision and affect both stress the particular features of the biological systems that implement them, the fifth chapter, on developmental neuroscience, goes deeper into the impact of the processes by which these systems emerge in humans. Some animals are "precocial"—relatively mobile and mature at birth—but humans are altricial, needing to be nurtured. As noted above, the human cortex is "experience ready," with some neuronal systems already adequate to begin the bootstrapping process of learning. Although the general pathways and broad functional organization of the cortex is genetically determined, much of the synaptic organization and the timing of its development are experience-dependent. Chapter Five gives an overview of the development processes, looks at the complex implications of the "experience-dependent" design, and concludes with an exploration of the dynamics of the emergence of the neuronal self.

The emergent neuronal self develops not just expectations but dispositions toward objects and events and towards its own body. The processes for memory—its encoding, maintenance, and appropriate recall—are crucial in the development of the neuronal networks of the brain and are the topic of Chapter Six. In modeling the important perceived regularities in the world, any change in synaptic weights in the construction of neural networks is a form of memory. However, most discussions of "memory" in the brain focus on higher order processes that mature as the brain develops in childhood. Traditionally, memory has been divided into a set of distinct functional types: two types of memory used in immediate processing—short-term and working memory—and two types of long-term memory: procedural and declarative memory (which then is divided into episodic and semantic memory). The chapter will focus on the development of episodic and semantic memory. It begins with the neuronal networks for initially encoding experience into episodic memory and then shifts to sleep and the process of memory consolidation which creates the links between episodic and semantic memory. The chapter concludes by briefly looking at the neuroscience of a particularly complex, flexible and distinctly human form of high-order combinatorial semantic memory structures—language—and what linguistic objects (texts) contribute to the shaping of the self as it is articulated within semantic memory.

The first six chapters examine key neuronal systems that provide the substructure for experience, selfhood and meaning. The final chapter uses two examples to explore the implications of the microstructural dynamics for the highly mediated patterns of lived experience. Since the neuroscientific models in this book analyze how we attend to the world, they serve as a form of phenomenology, which studies the logic of how we apprehend the objects and events that appear before us. In the first part of the chapter, I therefore consider analogies between the neuroscientific account and the formal phenomenology of Jean-Luc Marion as presented by Cassandra Falke's application of Marion's phenomenology to the experience of reading. I end with a comparison of the embodied, phenomenological model of meaning implicit in the neuroscientific framework with the classical Chinese poetic tradition, which is built on assumptions that mirror the neuroscientific model in significant ways. I offer these frameworks from very different traditions as two examples—among many possibilities within our broader humanistic endeavor—of ways in which we can contribute distinctive articulations of human experience both to our conversations with neuroscience and to our own explorations of how we, as biological creatures, participate in a world deeply resonant with meaning.

Chapter One

Reframing

Meaning lies in patterns of connection and their implications. Humans have evolved as creatures powerfully built for the extraction of—and response to—patterns in the world, in the body, and in the body's interaction with the world. This chapter introduces the microstructure of the human capacity to discover and create meaning and lays the foundation for all that follows in my account. The chapter introduces the basic models for neural network processing through which the brain organizes patterns, from basic perception to memory, emotion, and the very creation of the self. All forms of embodiment and construction of meaning in human experience—at least within our knowledge of the phenomenal realm—occur through the mediation of the system of neural networks in the brain. However, these neural networks, as biological systems, operate at a level very far below what we encounter in our usual understanding of human experience and, at the same time, work through rules profoundly different from the computer programming model that has become the analogy for more mechanistic accounts of the human. Understanding the basic processes of neural networks requires an imaginative as well as conceptual shift as we replace the apparently seamless immediacy of the surface of experience with the complex understructure of neural systems for constructing patterns that supports our sense of a world.

The chapter starts with neurons and their synapses, the basic elements of the brain's neural systems. Neurons by themselves simply "spike" (send patterns of electrochemical activation along their cell membrane): their capacity to produce all the effects we experience in mental life is through participation in neural networks. The central focus of the chapter, therefore, is on the behavior of networks of neurons: I begin with historically important simple models and the mathematics that support them and then introduce some of the ever more sophisticated networks through which researchers have explored the complex emergent properties of neural systems. The networks I present grow increasingly complex in their patterns of connections, from feed-forward systems to networks with added layers and with lateral and feedback architectures.

At the end of the chapter, after I have introduced models for fully articulated systems, I turn to the important contemporary paradigm of "predictive coding" that has grown out of several decades of slowly acquired understanding of the behavior of neural networks. The conceptual shift to thinking of our experience of the world as constructed out of systems of neuronal networks is difficult enough, but the predictive coding model asks us to take a second step and think of our experience of the present—all of our perceptual and conceptual apparatus—as never the present itself but as read through our neuronal structuring of the past.

I have tried to keep the details of the neuroscientific models at a level appropriate to the goals of joining in meaningful conversation with the neuroscientific community and of rethinking the conceptual framework with which we approach our own disciplines. Humanists, I think, need to have some idea of what neurons do, of what the mathematics behind neural networks looks like, of how learning and response in neural networks work, and of what the implications of predictive processing are for our experience of the world.[1] Understanding this conceptual language will both help us reimagine the human and share our own distinctive approaches to the human with our colleagues in the sciences.

---

[1] This is not a form of biological reductionism. While this book presents general principles that allow us to draw a few conclusions significant for humanistic studies, the biological systems that articulate these principles are of such vast scale and fundamental variability in their specific details that the very idea of "biological reductionism" in the study of human experience—the effort to reduce the individuated, historically shaped self to knowably deterministic biological processes—simply does not work.

Neurons and Neural Networks

NEURONS

Neurons are at the heart of our story because the mind, as is often noted, is a functional state of the brain, and that brain does what it does, has the characteristics that it has, through the behavior of the neurons out of which it is structured. Neurons transmit information from our sensory apparatus to the brain, within the brain, and from the brain to our muscles and other response systems. The basic process of neuronal communication begins with the reception of chemical signals at the dendrites of a neuron, the transmission of those signals along neuron's axon, and the re-transmission of the signals through bursts of chemicals released at the synapses linking the axon terminals of that neuron to the dendrites of the "receiving" neurons, which then repeat the process. Within the "sending" neuron itself, the transmitted information takes the form of patterns of "spikes" (more formally called "action potentials"), waves of depolarization that travel along the surface of the axon.

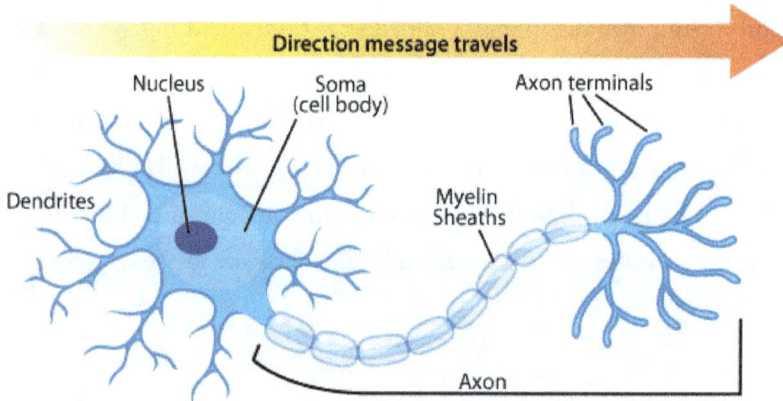

Figure 1.1: The Neuron[2]

The wave of depolarization, "spiking," locally reverses the difference in concentration of electrically charged ions between the inside and outside of the

---

[2] Drawing of Neuron courtesy of Brett Szymik. "A Nervous Journey." http://askabiologist.asu.edu/neuron-anatomy. © Arizona Board of Regents / ASU Ask A Biologist. Published under Creative Commons BY-SA 3.0.

cell. In neurons, the imbalance is a net *negative* charge inside the cell and a *positive* charge outside. The neuron creates this polarization by running an "ion pump" on the cell membrane that moves sodium (Na^+) ions *out* of the cell and moves potassium (K^+) *into* the cell. Because the potassium level is higher inside the neuron than outside *and* passes through the cell membrane relatively easily, the outward diffusion of the positive potassium ion creates a negative charge inside the cell. The level of polarization in the steady-state equilibrium between the activity of the ion pumps and the countermovement of diffusion determines the *resting potential*—a measure of the difference in electrical charge—of the cell membrane.

The two major ways *depolarization* is activated are via receptor potential and synaptic potential. In the first case, various sensory neurons react to physical stimuli (light, pressure, etc.) to open Na^+ channels (or close K^+ channels). Of greater concern to the story here are the changes in *synaptic potential* that arise when the *neurotransmitters* released at the axon terminals of the "sending" neuron bind to receptor proteins at the synaptic junction of the dendrites of the receiving neuron. Some neurotransmitters cause Na^+ (or calcium, Ca^{2+}) channels to open, letting Na^+ or Ca^{2+} ions flow in and contribute to depolarization: they are *excitatory*. Other neurotransmitters increase the permeability of the cell membrane to K^+, letting it flow out more quickly and thus causing *hyperpolarization*, an increase in the *membrane potential*. Both receptor and synaptic potentials are *additive*: pressing a sensory neuron a little causes a slight depolarization; pressing it more produces greater depolarization. Similarly, twice as many neurotransmitters binding to neighboring receptor sites will produce twice the effect.[3] If both an excitatory and an inhibitory transmitter bind in close proximity at roughly the same time, they cancel each other out.

Local depolarization of the cell membrane causes nearby Na^+ channels to open and increase the depolarization. If the depolarization reaches a certain threshold, it stimulates a cascade effect where more and more Na^+ channels open.

---

[3] This is only roughly true, and there are no doubt many variations. For example, there are bipolar cells in the brain that are tuned to detect simultaneous input at a given frequency from each ear. The response to simultaneous binding on each of the two dendritic branches is far greater than two signals on the same branch.

Na⁺ rushes in and produces a spike. This rapid local depolarization produces a well-defined change in voltage across the cell membrane called the *action potential* and strongly affects the nearby Na⁺ channels, causing them to reach the threshold and also spike. Since the action potential quickly dissipates as slower-acting K⁺ channels also open and counterbalance the Na⁺ influx, the net result is a wave of depolarizing action potential that travels down the cell membrane from the dendritic synapses to the axonal terminals and their synapses. The action potential spike causes the cell to release neurotransmitters into the axonal synaptic junctions, where they bind with the receptors on the dendrites of the next layer of neurons.

This, then, is the basic function of a neuron: to use chemical and electrochemical signaling to respond to the incoming patterns of activation from all the axons to which its dendrites have synaptic connections and produce outgoing patterns of chemical release. Underlying this function, however, is a great biological complexity that shapes the actual dynamics of response. This complexity is important in our thinking about ourselves as biological beings.

Beyond the Serial Computer Program: Neural Networks

The spiking of neurons that I have just described does not seem to do very much. How does one get from neuronal spiking to a mind, meaning, and a self? To begin to get an answer to this central question, we must turn from the activity of single neurons to properties that emerge from the vast, structured networks of neurons that form the brain. In the adult cortex there are between 12 and 20 billion neurons, and for each neuron there are about 7,000 synaptic junctions.[4] The average response time (between receiving the activation input at the dendrites and transmitting a pattern of activation to the axonal terminals) is about 5 milliseconds, while the average observable conscious response to a stimulus occurs in about 500 milliseconds. Thus the brain typically produces a response via about 100 steps of neuronal activation (i.e., 500 milliseconds / 5 milliseconds per step).

---

[4] Bente Pakkenberg et al., "Aging and the Human Neocortex," *Experimental Gerontology* 38 (2003) 95–99.

This illustrates why the "computer programming" analogy fails. The "one hundred step" process shows that whatever is going on to produce our observable responses from the basic logic of the firing of single neurons is not a form of computation that can usefully be compared to what we usually think of as a computer program. The computers with which we are familiar run in a serial fashion:

```
LanguageID = LanguageSettings.getLanguageID(system)
If LanguageID = 2052 Or LanguageID = 3076 Then
    gDisplayLanguage = "SimplifiedChinese"
Else
   If LanguageID = 4100 Or LanguageID = 1028 Then
      gDisplayLanguage = "TraditionalChinese"
   Else
      gDisplayLanguage = "English"
   End If
End If
Call ChangeLanguage(gDisplayLanguage)
```

They perform one step at a time following a program that provides logical conditions for executing instructions. Researchers have long understood that since one hundred instructions in a serial computer do not get one very far, the brain must use a mode of computation that relies on an almost inconceivably massive parallelism in its neuronal networks. This mode of computation differs profoundly from the step-by-step logic of a serial computer program.

Modeling the Neuron

Starting in the late 1940s, the challenge confronting researchers was to model what large systems of neurons could do. A fundamental part of all such models was proposed well before researchers began to use computers to simulate the behavior of neural networks. In 1949 Donald Hebb, a neuroscientist, published *The Organization of Behaviour: A Neuropsychological Theory*, in which he argued:

> When an axon of cell A is near enough to excite cell B and repeatedly or persistently takes part in firing [cell B], some growth process or metabolic change takes place in one or both cells such that A's efficiency, as one of the cells firing B, is increased.

Although his formulation is in descriptive terms, the researchers in the 1950s who developed the initial mathematical models turned it into the algorithm discussed on page 25. They began by exploring the behavior of networks of simple artificial neurons and then, once they had more experience, considered ways to shape that behavior. They began with computer simulations of networks made of artificial neurons, mathematical models of the simplest form of a generalized neuron:

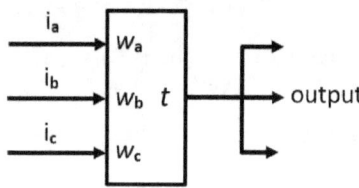

w_i is the **weight** (the *strength*) of the **synaptic connection** between the input connection **input i(i)** and the neuron. *T* is the *threshold* for the weighted sum of input activations at which the neuron fires.

output = 1, if $\sum [\text{input}(i) \times w_i] >= t$
          0, otherwise

In my illustration, the neuron has three dendritic synaptic connections to an input layer of neurons. Each synapse has a numerical weight assigned to it that corresponds to the strength of the synaptic connection in actual neurons, a weight that is determined by the number of receptors at the synapse as well as changes in the structure of the synapse and other factors that vary with the activity of the synapse. The "depolarization" of the neuron is the sum ($\sum$) of the spiking rates of all the input neurons times the weight of their synaptic connections. Each neuron also has a threshold value t for the total "depolarization:" if the sum is equal to or greater than t, then the neuron fires and passes this spiking information along its axon to three output synaptic connections. In the simplest model, the "spiking rate" is either just 1 ("it is active") or 0 ("it is not active").

Just as behavior of the single neuron—its capacity to spike—seems extremely limited, the single artificial neuron seems equally unpromising.

However, since it was clear to early researchers that the question was not so much what single neurons can do but what large populations of neurons connected to one another could do, they arranged the artificial neurons into simple two-layer networks. This is the classic design for the perceptron developed by Frank Rosenblatt in the late 1950s.

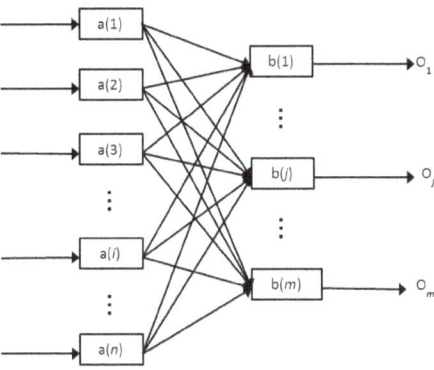

The first category of tasks to which such systems were assigned was pattern recognition. For example, consider a grid of 10 x 10 "sensory" neurons that turn on and off to represent simple graphic images of capital letters. The grid of neurons is to be trained to map the patterns of dots that represent the 26 capital letters onto a row of 5 neurons that represent the ASCII binary code for the letter ($2^5 =$ 32, which is more than enough to represent the 26 letters). From the perspective of the output neurons, the representations of the patterns for individual letters are not to be found in any particular sensory neuron but are instead in the pattern of connections within the entire network.

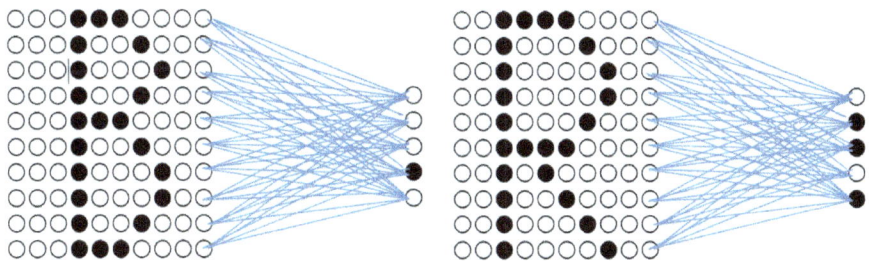

Figure 1.2: A Grid of Neurons to Recognize Letters
This diagram shows only the connections between the very last column of input neurons and the five output neurons

This group of 100 neurons "reads" the patterns for letters and represent those letters through their activation status (black = on = 1 or white = off = 0). These 100 neurons then pass their activation status to the group of five neurons that, as a group, were to give binary values for the capital letters ("A" = 00001, "B" = 00010, "C" = 00011,"... "Z" = 11010"). (The diagram, for simplicity's sake, shows the connection for just the last column of ten units.) Before training begins, the weights of the connections between the 100 sensor neurons and the 5 aggregating output neurons are randomly assigned, and the initial responses of the 100 neurons and the effect they have on the 5 output neurons is correspondingly random. However, a training rule for the five neurons is introduced: the output neuron fires if it receives "on" signals from 10 input neurons, and for any of the 100 neurons that contribute when an output neuron fires, the weight of the connection between those two neurons is increased. If one repeatedly trains the five neurons on the patterns for the 26 letters for thousands of training cycles and adjusts the weights according to the training rule each time, it turns out that those five neurons eventually learn to identify the 26 capital letters from the patterns of input from the 100 "sensory" neurons. This seems like a bit of magic, but it does work (for certain types of patterns). From trial to trial, the actual binary code assigned to each letter in the output *will be different because for each trial*, the initial synaptic weights for the 500 connections will be different and lead to different patterns of activation.

The Simplest Version: Hebbian Learning

Realizing that a simple network like the perceptron could be trained through repetition and adjustment of connection weights to extract patterns was an important step. The equation used to train the network is what has come to be the mathematical version of Hebb's learning rule:

$$\Delta w_{ij} = \eta a_i b_j$$

That is:

a_i and b_j = the output values for the ith unit in Layer A (here, with 100 neurons) and the jth unit in Layer B (here, with 5 neurons)

Δw_{ij} in = the change (Δ) in the *connection weight* between the ith unit in Layer A and the jth unit in Layer B

η = a learning rate parameter

Note that if either a_i or b_j is 0 (i.e., the neuron did not fire), then the connection weight does not change. The weight changes only if both the input neuron and the output neuron fire. The idea is that when the output neuron fires, one should strengthen the connections of all the input neurons that contributed to the firing. The popular shorthand explanation of Hebb's rule has thus become "Neurons that fire together wire together."

The system of weights from the 100 input neurons to the 5 output neurons that are adjusted but eventually reach a stable solution can be represented mathematically. The weights for the connections from the 100 sensor neurons to each of the output neurons can be listed as a row of 100 values. There are thus 5 rows. This collection of all the values of the synaptic weights between the 100 input neurons and the 5 output neurons defines a weighting matrix:

$$\mathbf{W}_{100,5} = \begin{bmatrix} w_{1,1} & w_{2,1} & \cdots & w_{100,1} \\ w_{1,2} & w_{2,2} & \cdots & w_{100,2} \\ \vdots & \vdots & \ddots & \vdots \\ w_{1,5} & w_{2,5} & \cdots & w_{100,5} \end{bmatrix}$$

A matrix in its definition is nothing more than this, an orderly list of the values of the weights. But matrices have very useful and well-studied mathematical properties that greatly assist in the analysis and modeling of neural networks. A second mathematical construct used in modeling neural networks derives from thinking of the list of the activation states (either 0 or 1) of all the input sensor neurons as a vector $\vec{\iota}$ with 100 elements that describes the total activation state of the input layer. Similarly, the output activation states of the 5 output neurons can be represented by a vector $\vec{o}$ with 5 elements. The weighting matrix and activation state vectors are at the center of our concern in the study of neural networks and provide powerful tools for representing and exploring the behavior of neural networks. "Learning" is the changing of the individual values in the matrix. The matrix itself is a form of memory that stores values and, at the same time, we can think of it as the "program" that maps the input vector $\vec{\iota}$ (a collection of 100 values that defines the input state of the system) onto an activation-state vector $\vec{a}$ for the five output units. One simply multiplies the input vector times the matrix:

$$\vec{i} \times \mathbf{W} = \vec{a}$$

Then there is an "activation function" (that contains information about the threshold for firing for the output neurons) that determines whether the five output neurons fire:

$$f(\vec{a}) = \vec{o}$$

This is a two-step calculation that relies on the array of 500 connections in the network represented by the weighting matrix **W**. The weighting matrix **W** and its use thus point to the *distributed* nature of representation and processing in neural networks.

Box 1: Mathematical Notation and the Real World

This chapter uses a few formalisms from mathematics, but it is important to see that there is nothing exotic in this standard notation. The two simplest are:

Δx "delta x" means simply the change in some value (usually over time), $x_{t2} - x_{t1}$
$\sum x_i$ "sigma notation" the sum of a series of values, $x_1 + x_2 + x_3 + ... + x_n$

Change in water usage from month to month, change in the value of a house, and change in the cost of apples all can be describes through the shorthand notation Δusage, Δvalue, Δcost.

Similarly, sigma notation is a convenient way to represent, say, the total water usage over a year: $\sum \text{usage}_{monthly}$. Both types of notation are yet more convenient in more complicated situations where one performs the exact same action to each item in a large collection.

Thus, in the chapter, we have large collections of neurons and their synapses, all of which follow simple, uniform rules:

$\sum [\text{input}(i) \times w_i]$ refers to the sum of (each of the input values for a neuron's synapses multiplied by the weight of the input strength for that particular synapse)

Δw_{ij} refers to the change in the weight (strength) of the synaptic connection between a neuron in one layer and another neuron in the next layer from one time cycle to the next.

Box 2: Vectors and Matrices

We live in a multidimensional world. To describe mathematically what happens in our multidimensional world, we need multidimensional "numbers." For example, to describe the position of any object in three-dimensional space, we need a set of three numbers:

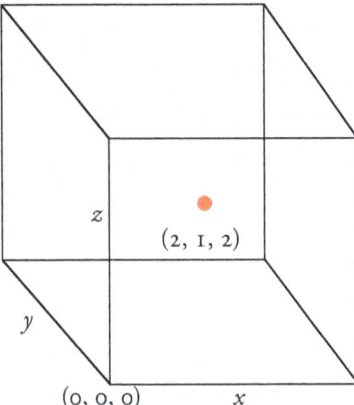

We can think of this point (2, 1, 2) as a displacement (shift) from the origin (0, 0, 0) along each of the x, y, and z axes. This displacement, represented by the three numbers, is a **vector**. The one-dimensional numbers, in contrast, are **scalars**. Vectors, as multidimensional numbers, have both a length (in this case, 3) and a direction.

In general, any set of n scalars can be described as an n-dimensional vector defining a position in an n-dimensional space. This is mathematical notation, but what it describes can be very real-world thing. The position of a boat on the ocean, its speed, and its acceleration all are three-dimensional vectors.

In our case, the *connectivity* of a neuron with a thousand synapses with a thousand corresponding connection weights can be described with a 1000-dimensional vector.

A **matrix**, as a mathematical formulation, is an expansion of a vector: instead of a single column of numbers (an $m \times 1$ matrix, with m rows and one column), it is a large set of columns of numbers (an $m \times n$ matrix, with m rows and n columns). However, what *is* a matrix as a thing? I suggest that, for our purposes, it can be thought of as a defined *way to transform vectors, to map vectors from one space onto vectors in another space*. Matrices are mostly used to analyze complex situations in economics, geology, topic modeling in digital humanities and behind the scenes in computer graphic animations. Generating how the moon reflects off of waves of water in a computer animation, for example, involves rays of light of varying brightness at a particular place arriving at a particular (3-D) angle, which will be reflected based on the angle of the water in a location at a particular time. The input and output are both vectors (rays of light), while the computer generates a dynamic matrix that describes the surging surface of the water. The matrix here mathematically represents a complex *thing* through its rows and columns of numbers: it describes waves as a surface with concavities and convexities. For a neuron network, the weighting matrix similarly describes a form of "energy" surface with local concavities of low energy that serve as attractor basins, discussed below. For each neuron, interacting with this surface transforms input activation vectors into output activation vectors.

Building models of neural networks is very much a cumulative project in which learning the behavior of simple systems helps prepare for the exploration of increasingly complex models that better represent the neuronal systems of the brain. Researchers worked with the formalisms of the perceptron—weighting matrices, activation vectors, and Hebb's learning rule—but were aware both that the connectivity in the brain was far more complex and that, as discussed below, the mathematics of two-layer feedforward perceptrons contained a seemingly fatal flaw. Interest therefore moved beyond the perceptron architecture as scientists added new network elements both to introduce features from biological systems and to overcome the constraints of the perceptron. For example, GABA (*gamma*-aminobutyric acid) is, in most cases, an inhibitory neurotransmitter, so scientists created "winner-takes-all" models where at the output layer the first neuron to reach the activation threshold inhibits all other neurons in the layer. Consider another small network designed to learn a simple pattern:

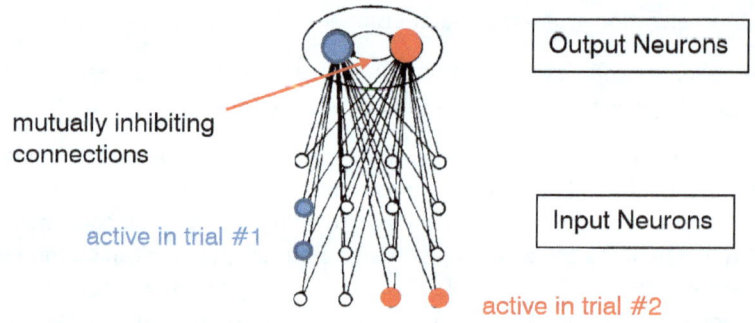

Figure 1.3a: "Winner takes all" Neural Network

In this system, only two neurons in the first layer fire at a time, and they must be next to each other either horizontally or vertically. In the second layer, only one of the two neurons can fire at a time (they inhibit each other: the one that fires keeps the other from firing).[5] When each trial begins, the weights of the connections between the neurons in the first layer and those in the second are

---

[5] This example comes from D. E. Rumelhart and D. Zipser, "Feature Discovery by Competitive Learning," in David E. Rumelhart and James L. McClelland, and PDP Research Group, *Parallel Distributed Processing: Explorations in the Microstructure of Cognition, Volume 1: Foundations* (Cambridge, MA: MIT Press, 1986), pp. 170-177.

given random values and then are trained through many rounds of activation of pairs of neurons in the first layer. Because the two output units are in a competition and because the weights for connections between input and output neurons are based on the successful activation of the output neuron, the learning process here is described as "competitive learning:" only the winner updates its connection weights. In different trials, the results were:

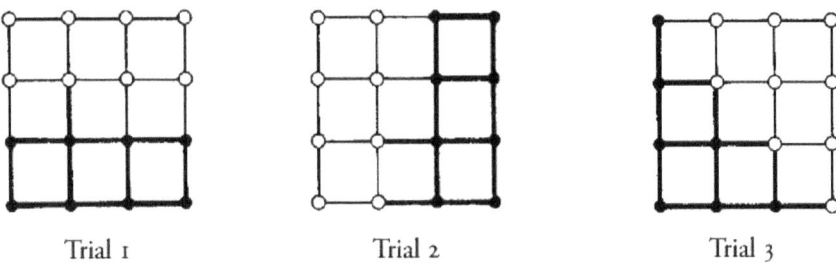

Figure 1.3b: "Winner takes all" Neural Network Trial Results

Filled circle: Output Neuron 1 gave the input from these units a higher weight,
Empty circle: Output Neuron 2 gave the input from these units a higher weight.
Heavy line: When the two input units connected by heavy lines were active,
               Output Neuron 1 becomes active
Thin line: When the two input units connected by thin lines were active,
               Output Neuron 2 became active.

At the end of each trial, when the system of weights and the pattern of activation have become relatively stable, the neural network has discovered a way to divide the input layer into two halves (i.e., it has discovered a topographic map), but in different trials, the system finds different solutions based on the initial weights. That is, since the output has only two states, the system must devise a way to partition the input domain (the sixteen neurons) into two units. Since, moreover, the rule for the game always requires that active neurons be adjacent, the solutions naturally will be topographical.

The line segment parser is a very simple example, but it resembles the way in which the "simple cells" of the primary visual cortex (V1) learn to respond to the input from the "On" and "Off" neurons of the lateral geniculate nucleus (LGN) of the thalamus:[6]

---

[6] For an explanation of the LGN, see the discussion of the visual system in Chapter 3.

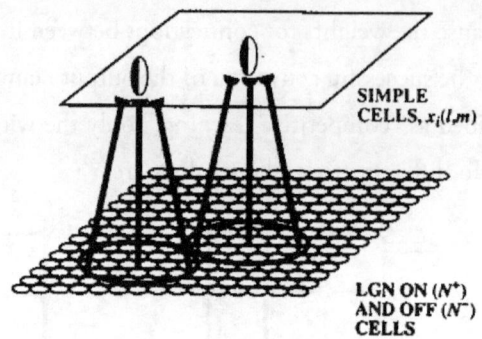

Figure 1.4: Model of Neural Network connecting the Lateral Geniculate Nucleus to the "Simple Cells" in the Primary Visual Cortex[7]

The cells in the LGN respond to whether, in a cluster of neurons in the retina, the central neuron is firing while those surrounding it are off ("On" cell) or if the central neuron is off and those surrounding it are firing ("Off" cell). The "simple cells" in V1 in turn aggregate the input from adjacent LGN neurons in a manner similar to the line-segment processor above and, through Hebbian learning, learn to respond to short line segments at different angles.[8]

Hidden Units

The perceptron's Hebbian bottom-up, feed-forward learning based on a relatively simple model of layers of neurons can explain important features of some systems in the brain and demonstrates the pattern-extraction power of massive networks of neurons. While the neuronal networks of the brain surely function something like this, the perceptron itself remains rather limited in its ability to organize synaptic connections. An actual brain employs a wide range of other strategies as it learns to connect complex internal needs to the shifting patterns it detects in

---

[7] This model comes from Steven J. Olson and Stephen Grossberg, "A neural network model for the development of simple and complex receptive fields within cortical maps of orientation and ocular dominance," *Neural Networks* 11 (1998):189-208.

[8] The "complex" cells in V1 learn to detect additional features like movement. See Chapter 2.

the sensory realm based on data from both the external world and from the bodily milieu.

Perceptrons also have a deep flaw at a basic mathematical, conceptual level that slowed the early exploration of neural networks until researchers found ways to train so-called hidden units to overcome the encountered barrier. In 1969 Marvin Minsky and Seymour Papert demonstrated that there are combinations of input patterns and output patterns that represent crucially important logical operations but that cannot be learned by two-layered perceptron systems.[9] For a period after the publication of Minsky and Papert's paper, the focus of research in artificial intelligence largely shifted to the construction of expert systems based on serial computational methods.

However, the same paper by Minsky and Papert that presented the limits of perceptrons also noted that systems with an additional layer of units in between the input and output layers *could* learn to produce the desired pattern response. These neurons that have no contact with the external world (i.e. the world of measurable input and output) are called "hidden units."

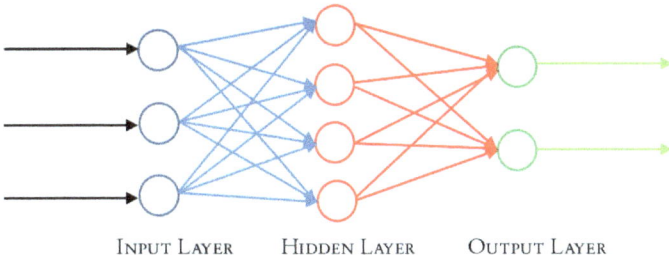

INPUT LAYER   HIDDEN LAYER   OUTPUT LAYER

---

[9] One of these is the so-called XOR ("exclusive or") where if $I_1 = 1$ (active) and $I_2 = 0$ (inactive) or $I_1 = 0$ and $I_2 = 1$, then the output neuron should be active, but if $I_1 = I_2 = 0$ or $I_1 = I_2 = 1$, then the output neuron should be 0. No combination of weights connecting the input neurons to the output neuron multiplied by the activation state of the input neurons can reproduce this result. Since the XOR is a fundamental logical element in serial computer design, this conclusion underscored the limitation of simple perceptron systems to researchers at the time.

The hidden units make neural networks far more powerful and flexible, but there was no easy way to train them to produce the desired output patterns mapped to a set of input patterns.[10]

Thus the matter stood for a decade. However, during this time, researchers continued to explore the behavior of the more complex forms of neural networks with hidden units, since this clearly was how the brain had to be working, and the question was how neural networks learned, as they manifestly did in the brain. The major breakthrough came from the development of a variety of strategies for using the difference between the desired and the observed activations in each layer to send instructions back to the neurons in the layer below it about how they needed to change their connection weights to get results that were closer to producing the desired outcome. This approach to training a network is called supervised learning, in that the desired outcome is known, and the programmer running the simulation changes weights between neurons to move the results in the right direction.[11] This training strategy of starting with the output layer and working backwards, layer by layer, for systems with hidden units is called the *back-propagation of error*.

The algorithms used to calculate the adjustments to the weights and even the pattern of bidirectional connectivity that these systems use are wildly non-

---

[10] Consider, for example, a solution to the XOR problem. The hidden unit is strongly *inhibitory*: its connection weight to the output unit is -2, but it fires only when its threshold of 1.5 is reached (which happens only when both input units fire)

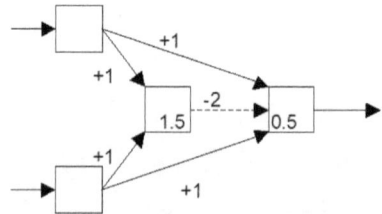

Output Unit activation: (0,0) 1 x 0 + 1 x 0 + -2 x 0 = 0
                                    (1,0) 1 x 1 + 1 x 0 + -2 x 0 = 1
                                    (0,1) 0 x 0 + 1 x 1 + -2 x 0 = 1
                                    (1,1) 1 x 1 + 1 x 1 + -2 x 1 = 0

[11] Hebbian learning, in contrast is *unsupervised*: the system simply settles into whatever patterns it finds in the input data.

biological. The researchers were perfectly well aware of the non-biological character of their solutions, but the crucial point was that the systems worked, and, as people gained ever more experience with these training techniques, the neural networks constructed through them became ever more powerful. In recent years, the striking success of so-called "deep learning" networks has achieved public visibility: these systems, relying on training through back-propagation of error, have beaten humans at Jeopardy and defeated the best players in the world at the game of "Go." Much of the current work in artificial intelligence (AI) explores ways of extending and integrating the capabilities of deep learning networks in different domains. The fact that these networks use approaches to learning that do not correspond easily to the way the brain functions is of little concern. The researchers want to build powerful pattern recognition systems that can plan actions (drive cars, bomb enemy tanks, purchase stocks) based on predictions drawn from past and present circumstances. While much of this research is driven by the need to find solutions for practical problems, the emergent properties of the systems that AI researchers develop remain important for neuroscience because they give researchers concerned with biologically plausible designs ideas about complex network behavior which they can apply to their search for analogous biological networks. That is, neuroscientists know that the neuronal networks of the brain rely on hidden units that stand between input and output layers and that the architecture of the brain must accomplish the same tasks as those performed through the back-propagation of error. Experience with these non-biological systems is profoundly helpful in thinking through the role of the feedback connectivity ubiquitous in the brain.

Networks with More Complex Connectivity

Simple Hebbian perceptrons illustrate how neural networks function as pattern detectors as they learn to attune to significant regularities in an input domain. The complexity of the patterns that can be mapped is expanded by the addition of hidden units and by the back-propagation of error that provided an initial strategy for training those hidden units. However, since back-propagation of error is so clearly a non-biological way to train networks, researchers have continued to explore other, more directly biologically inspired models for constructing networks that use more complex designs and flows of information.

Self-organizing Maps

Among biological neural networks, neuroscientists have long focused on modeling the neural networks that underlie the visual system and, in particular, on the initial transmission of spiking information from the retina to the lateral geniculate nucleus (LGN) of the thalamus and then on to the primary visual cortex (V1). The basic logic and functionality of this system are sufficiently simple that it provides a good basis for experimenting with neural network design. The primary visual cortex translates the information from the LGN into patterns of activations of short line segments at different angles organized in a way that corresponds to the spatial organization of the retina itself. (That is, V1 is *retinotopic*.) For each location, neurons are trained to respond maximally to line segments of particular angles. The neurons thus become feature detectors. Initially, researchers focused on modeling the interconnected system of neural units that produced the characteristic response curves for line segments in V1.

Neural networks built on the model of the internal connections within V1 give us deeper understanding the general functioning of the neuronal networks in the brain. Drawing on the organization of V1, researchers discovered that modeling the *lateral connectivity* between neurons *within* a layer makes possible an extension of the basic function of unsupervised Hebbian learning in the construction of feature detectors that behave like the columns of neurons in V1. Computer simulations showed that layers of neurons with intra-layer connections create *self-organizing maps*: they map out the structure of the domain of the input they receive. These models are an example of how computer simulations have given researchers insight into the biological systems that they model. Neuroscientists have come to understand that for cortical networks that receive input from neurons that translate external sensory data—like V1 for the regions of the retina (via the LGN) or the primary sensory cortices that process touch—these neuronal self-organizing maps identify structured features of the input in a topological manner. The set of mutually differentiated features developed in any one localized network is not very complex (e.g., short line segments at different angles in V1), but V1 then passes its first-order structuring of the data on to a next area that develops a mapping of these first-order features into yet higher order structures (sets of recurring shapes in V2 and V3). These localized networks in turn pass their sets of more highly structured data to networks that continue the process of abstraction to create increasingly higher dimensional representational maps. That is, while there are differences in the structure of various regions of the cortex, researchers who model the neural networks of the brain—and the researchers who have come to rely on these models as adequate representations of neural systems in the brain—have become confident that the construction of self-organizing maps is the *basic* task of the neural processing of input, from the primary sensory cortices through to the areas that embody semantic memory and the planning of action.

The simplest model for self-organizing maps (SOM) draws on the intra-layer interconnection of neurons and feed-forward connections to higher order processing areas.[12]

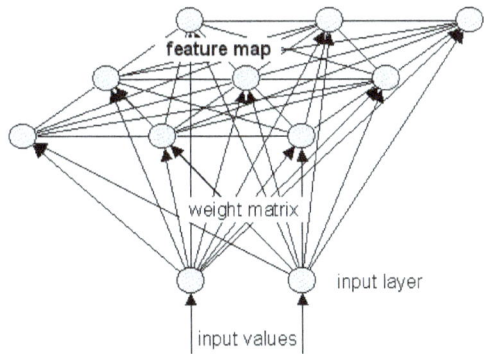

Figure 1.5: Kohonen Self-organizing Map Network[13]

---

[12] Chapter Two on the visual system provides additional details on aspects of these self-organizing maps. Those interested in a more formal presentation of the mathematics can consult Thomas P. Trappenberg, *Fundamentals of Computational Neuroscience* Second edition (Oxford: Oxford University Press, 2010), pp. 183-212,

[13] Accessed from Jochen Fröhlich, Neural Networks with Java, https://www.nnwj.de/kohonen-feature-map.html on November 18, 2021.

The Kohonen network, the most famous SOM, uses a winner-take-all approach in the learning phase. That is, the feature unit (for angled line-segments in V1, for example) whose weights for the input units most closely matches the input vector dominates the pattern for learning from the input. Its connection to active input neurons increases the most; its nearest neighbors' connection strengths increase a bit, while those at a certain distance are decreased. This type of learning remains essentially Hebbian, even if the update spreads to neighboring units. The next step taken by researchers in exploring SOM is a variation on Kohonen's model—the *laterally interconnected synergetically self-organizing map* (LISSOM) —in which each unit derives both its activation and the change in its weight vector from the effect of the input (afferent) signals and the total effect of all its neighbors. Following the approach of Kohonen's initial SOM design, the weights of the connections between the interconnected neurons within the feature-map layer rapidly diminish with distance. Neighbors influence one another in a positive, excitatory manner but, at a certain distance, the connections become inhibitory. The graph for this function often is called the "Mexican hat." Because the connections of the afferent (input) layer are sparse—that is, each input neuron is connected to only a subset of the neurons in the feature-extracting layer—and because in V1 the connections are spatially arranged, the result is a topologically organized feature-detecting network where the same features are identified for different, spatially separated clusters of neurons.

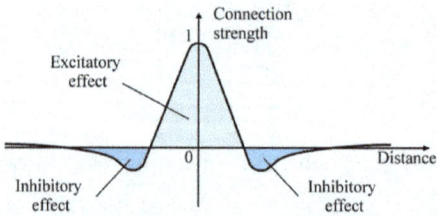

Figure 1.6: "Mexican Hat" Weight function for Lateral Connections in SOM[14]

---

[14] A. Mwegerano, P. Kytösaho and A. Tuominen, "Applying Self-Organizing Maps Method to Analyze the Corrective Action's Quality Provided to Customers with Mobile Terminals," *iBusiness*, Vol. 4 No. 2, 2012, pp. 108-120. doi: 10.4236/ib.2012.42013

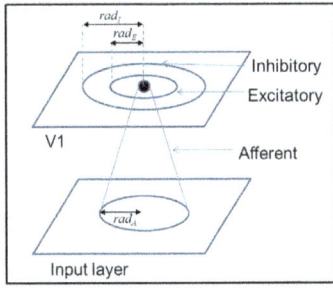

Figure 1.7a: Architecture of the LISSOM model[15]

Figure 1.7b: LISSOM Topological Feature Map Model for Line Orientation in Hypercolumns in V1[16]

It turns out that while the initial models were based on the visual system, these models in fact can be much more robust in their application. James Bednar, exploring how well LISSOM models can be used to recreate the observed features of the primary visual cortex, concluded that there was nothing inherently "visual" about feature-extracting, topologically organized, laterally connected self-organizing maps: "Because the model cortex starts without any specialization for vision, it represents a general model for any cortical region"[17]

---

[15] Philips RT and Chakravarthy VS, "A Global Orientation Map in the Primary Visual Cortex (V1): Could a Self-Organizing Model Reveal Its Hidden Bias?" *Frontiers in Neural Circuits* 10.109 (January 2017), p. 4.

[16] The different colors are for different angles. The figure is from Philips and Chakravarthy, "A Global Orientation Map in the Primary Visual Cortex (V1)" *Frontiers in Neural Circuits*, p. 11.

[17] J.A. Bednar, "Building a mechanistic model of the development and function of the primary visual cortex" *Journal of Physiology - Paris* 106 (2012) 209.

Recurrent Neural Networks

Simple perceptrons, multi-layer perceptrons (MLP) trained by back-propagation of error, and LISSOM feature-detecting networks all are types of *feed-forward* networks. Even though the training of MLP requires that information flow backwards, the model of connectivity is in just one direction. (The programmer acts as a deus ex machina who calculates and delivers the correction values from outside the network.) Researchers have long been aware, however, both that the brain must solve the problem of training hidden units through some form of feedback and that the brain does in fact rely on massive systems of top-down connections.[18] The challenge has been to model the functioning of those feedback connections. What is the role of the ubiquitous top-down, recurrent connections from higher-order to lower-order layers?

In the last two decades, researchers who model neural networks with feedback connections have been in constant conversation with neuroscientists who explore similar systems at the biological level. This highly fruitful collaboration has pointed to four basic functions for recurrent connection in biological neural networks. The first two are closely connected: the resolution of ambiguous data and the filling out of incomplete patterns. Because the activation of higher-order units is based on abstracting patterns from lower-level units, when the higher-level units send patterns of activation back downstream, these are the "best guesses" about the initial lower-level input. However, the input data often is not very good, either because it is noisy or because it is just a partial sampling. The initial guesses transmitted by the recurrent connections nonetheless nudge the activation of those input units in a particular direction—towards particular, already-learned patterns of activation—which will in turn move the input units to send up patterns that yet more closely correspond to the learned patterns. In situations with incomplete data, the recurrent connectivity

---

[18] While the initial training of the visual system in early infancy follows the simple Hebbian rules of self-organizing maps, the story is actually more complex because the patterns of activation in processing the data from the retinas not only flow upward (retina → LGN → V_1 → V_2 ....) but also downward from V_2 to V_1 and from V_1 to the lateral geniculate nucleus. In fact, there are ten times as many recurrent connections from V_1 to the LGN than there are feedforward connections from the thalamus to the primary visual cortex.

can effectively push the network to fill in the missing information. Completing these two types of constrained data is a powerful and important feature of recurrent networks found in both the computational models and their biological counterparts.

A third function is crucial for the processing of patterns that unfold over time in a sequential manner: while there are many variations on the architecture in the computer modeling, in the simplest version, an input level will send output to a layer that serves as a "state machine" that preserves the former activation state and returns this input to the input layer to allow it to compare its former state with successive input patterns. The iterative nature of recurrent networks—the fact that recurrently connected neurons invariably are returning responses based on earlier states in both the simulations and the biological systems—plays a crucial role in a variety of models that seek to develop biologically plausible mechanisms for training complex cortical capabilities.

These three functions of recurrent connection in the brain—ambiguity resolution, pattern completion, and processing sequential patterns—have inspired significant work in both artificial intelligence and network neuroscience, but a final function for such recurrent connectivity is seemingly distinctive to the brain: its role in imaginative reconstruction and conscious awareness.

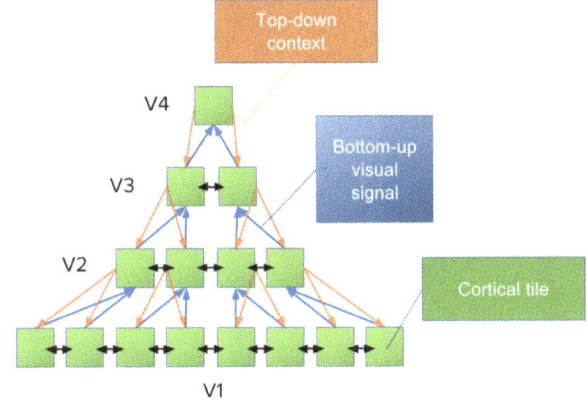

Figure 1.8: Model of Recurrent Connectivity in the Visual Cortices[19]

---

[19] Micah Richert, Dimitry Fisher, Filip Piekniewski, Eugene M. Izhikevich, and Todd L. Hylton, "Fundamental principles of cortical computation: unsupervised learning with prediction, compression and feedback" arXiv:1608.06277 [cs.CV] (August 2016), p. 6.

Attractor Basins

The inspiration for work on biologically plausible recurrent neural network models comes from the ubiquity of top-down connectivity in the cortex. The best-known system for experimenting with recurrent approaches is, once again, the highly recurrent visual cortex, whose design and functionality have been intensely studied for decades. Figure 1.8 is the representation of one artificial neural network modeled on the visual system. The process is a recursive one: each level receives the patterns extracted from the input given to the layer below it and aggregates those patterns into yet higher order ensembles of patterns. Then, in the next iteration, the layer *returns* activations based on its extracted patterns back to the neurons of the layer below it to serve as additional information for that lower level to use in identifying the most strongly salient patterns, which the neurons then send back up the pathways.[20] These recursive processes lead to a top-down refinement of the patterning. As the system matures and connection weights settle, each layer develops a way of partitioning the input space into a matrix of mutually differentiated *attractor basins*. When the initial input pattern falls within the shallow slopes of the attractor basin, the activation returned to the initial input layer will help drive the next set of input activations into a pattern yet more closely aligned with that defined by the attractor state. When this happens, the activation pattern converges on that of the attractor state within several iterations.

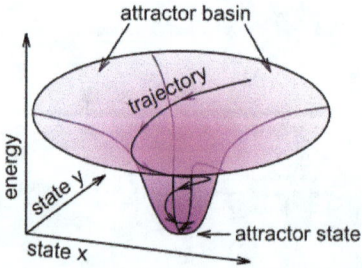

Figure 1.9: Attractor Basin[21]

---

[20] Micah Richert et al., "Fundamental principles of cortical computation: unsupervised learning with prediction, compression and feedback."

[21] The figure is from O'Reilly, R. C., et al., *Computational Cognitive Neuroscience*. Wiki Book, 4th Edition. URL: https://CompCogNeuro.org, Figure 3.15, p. 55.

Perhaps the best-know example of the power of top-down attractor dynamics to shape lower-level neural activation is the figure-ground separation in a high-contrast photo:[22]

Figure 1.10: Figure-ground Discrimination

High-level knowledge of "Dalmatian" quickly allows the visual system to identify the series of black spots as the figure of a Dalmatian set off against a background.

This general model of recurrent networks that develop attractor basin behavior has been long understood, as had a key problem with this model. How can such systems learn to recognize new entities that would require the restructuring of the entire map defined by mutually differentiated attractor basins? If a toddler has known only cats but is introduced to a toy poodle, its strange behavior—tail wagging, barking, face-licking—and peculiar curly fur would need an entirely new attractor basin to categorize the new word and experience of "dog" as separate from "cat."

If the toddler's system for categorizing sensory data, through its iterations of bottom-up and top-down looping of activation, always pulls the input data for small, furry four-legged creatures into its currently available, closest-fit attractor basin of "cat," how can it change? Stephen Grossberg and Gail Carpenter proposed a general solution in their *adaptive resonance model* (ARM),

---

[22] The figure is from Richard L. Gregory, "The Medawar Lecture 2001, Knowledge for vision: vision for knowledge," *Philosophical Transactions of the Royal Society London B Biological Science* 360 .1458 (2005), p. 1238.

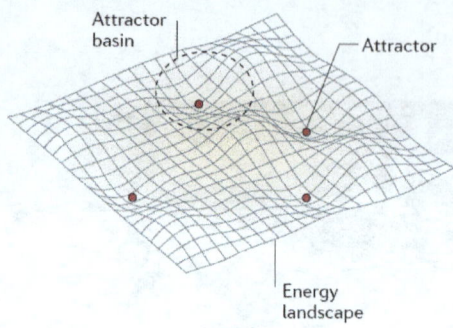

Figure 1.11: Attractor Basin Landscape[23]

which has gone through many versions as it has been applied to an increasing number of brain systems. The first processing layer (the "input/comparison field) extracts patterns from the input that it then transmits to the next layer (the output/recognition field) for yet higher-order pattern extraction (See Figure 1.12.).

Several cats in the house allow "cat" patterns to be extracted as a predictable sensory experience. The higher-order layer generates its own activetions that it simultaneously both sends onward and sends back to the lower layer. If the feedback pattern matches the initial feedforward pattern well enough ("It's another cat."), then the process of resonant looping continues in the usual manner. However, if the mismatch is great enough, as determined by the "vigilance test" unit ("Parents provide the word 'dog' for this strange-appearing, strangely acting "cat"), an inhibition of the most active neurons in the higher-order level is triggered and initiates a learning process that reconfigures the structure of the output/recognition field. Chapter Six on memory will return to this issue and to the biological systems that may implement the sort of reset process that Grossberg and Carpenter propose.

---

[23] See the discussion in Rafael Yuste, "From the neuron doctrine to neural networks," *Nature Reviews: Neuroscience* 16 (August 2015), p. 492. Yuste adapted the figure from Chris Eliasmith, "Attractor Networks" in *Scholarpedia* 2(10):1380. (http://www.scholarpedia.org/article/ Attractor network copyright Creative Commons Attribution-NonCommercial-ShareAlike 3.0 Unported License).

Reframing 39

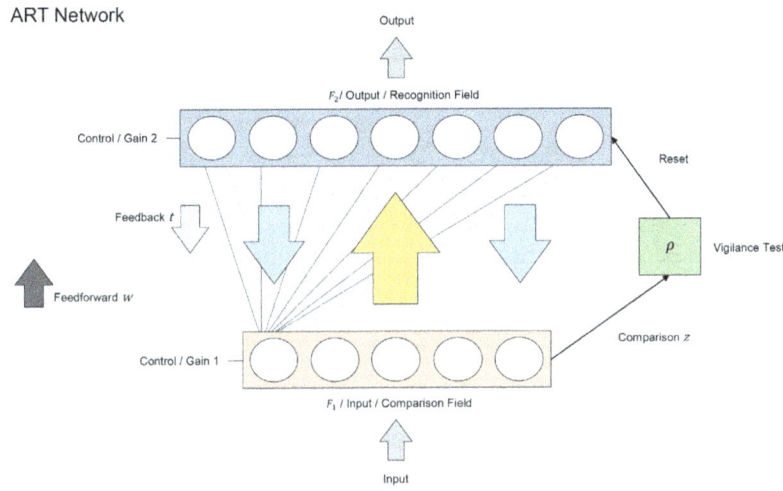

Figure 1.12: The Adaptive Resonance Model[24]

An Aside on the Mathematics of Spiking Neurons

The mathematics for the network models that I introduced at the beginning of the chapter rely on a simple step-function activation equation and on matrices for the strength of synaptic connections both in the neurons' learning and in their responses to input. The mathematics of contemporary modeling for LISSOM, recurrent networks, and ARM, in contrast, rely on complex sets of equations to more closely reflect the dynamics of neuronal network behavior. These models incorporate the fact that the brain's neurons convey their information as spikes of electrochemical depolarization when they reach their activation threshold: contemporary models use a neuron's spiking rate to represent its activation level and serve as the synaptic input and output in the network. Louis Lapicque long ago (1907) provided a "leaky integrate-and-fire" equation to model the spiking behavior of neurons that took into account both the ideas of summing synaptic input and firing on reaching a threshold and the neuron's biological behavior of constantly leaking charge. Modern computational models continue to update Lapicque's approach and use increasingly complex differential equations with ever more parameters to capture the actual behavior

---

[24] This image comes from https://chrismcbain.files.wordpress.com/2015/06/art4.png.

of the spiking of individual neurons. This spiking, however, not only has a rate and a duration, but also a phase relation with other spiking neurons providing input to a neural network. It turns out that while spiking rate is significant in networks of spiking neurons, so too are issues of synchronization of spiking activity. As one researcher explains these features:

> A convenient way to characterize neuronal signals is to consider them as a volley of spikes (a pulse packet) that can be quantified by the number of spikes in the volley (α, 50–100 spikes) and their temporal dispersion (σ, ~1–10 ms), which measures the degree of synchronization of the message. Several studies have shown that both α and σ are necessary to understand the downstream effect of a pulse packet.... Note that a pulse packet by itself does not carry information; rather, the code resides in the combination of neurons participating in the spike volley in the sender and receiver networks.[25]

The mathematics of these networks is now well developed, and researchers increasingly take advantage of the synergy between experiments that explore the behavior of particular types of networks and the running of simulations to test conjectures about that behavior.[26] While humanists, in reading neuroscience research articles, need not fully follow details of the mathematical arguments that are made, it is important to understand the conceptual framing of the discussions that are embodied in the mathematical modeling (which often are relegated to the "Methods" section in any case).

The Predictive Brain

One theoretical framework that has developed over the past twenty years and now plays a significant role in thinking about neural networks in the brain is the paradigm of "predictive processing." Neuroscientists, like other biologists, in general accept uncontroversially that the goal for all animals is to preserve

---

[25] Gerald Hahn, et al., "Portraits of communication in neuronal networks" *Nature Reviews Neuroscience* 20 (February 2019) p. 118.

[26] Contemporary researchers routinely test their experimental results against models constructed from standard neural networks software packages.

homeostasis, an internal state of the body that maintains appropriate conditions for the animal to function and survive. Under this assumption. the brain's task is to shape responses to the environment to preserve homeostasis. For simple nervous systems with little memory, the brain controls movement away from danger and toward food, etc. More complicated brains that can preserve memories develop models of the world to anticipate both bad and good future outcomes. Karl Friston proposed his "free-energy principle" as a formalization of the goals of predictive processing: the imperative to "avoid surprises" in assessing the current state of the world to which the brain must respond.[27]

Predictive Processing

It is crucial to remember that the neural networks of the brain know nothing about the external world or the body: they simply model the patterns of input that they receive and constantly face the key problem: how to avoid guessing wrong about the world. The model of the world and the body that the ensemble of neural networks in the brain develops finds larger structures in the data accumulated over time. At any particular moment, however, the brain draws on these developed patterns to both interpret and shape (via incomplete pattern recognition and ambiguity resolution) the current state of input from the body and the world. Friston's FEP serves as a basis for mathematical models for how predictive processing minimizes mistakes about the world. The framework in which "FEP fundamentally analyses existence in terms of the probability of finding the system in certain states and the corresponding *surprise* of finding it in others" points to computational processes of Bayesian inference, in which one

---

[27] Karl Friston's "free energy principle" (FEP), derived from thermodynamics via information theory, provides an important formal account of the logic behind predictive processing:

> FEP says that any self-organising system that exists (in the sense of being at non-equilibrium steady-state with its environment) must minimise its (variational) free energy. FEP fundamentally analyses existence in terms of the probability of finding the system in certain states and the corresponding *surprise* of finding it in others. This leads to the imperative to minimise surprise, for which free energy minimisation is required.

Jakob Hohwy, "New directions in predictive processing" *Mind and Language* 35 (2020), p. 213.

calculates the probability that an already-known state of the world considered "prior" knowledge could have produced the currently encountered state of the inputs.[28]

In the predictive processing model, if the most likely "prior" was based on input that—in previous instances—looked a bit different from the current version, the network sends the prior pattern as feedback to the lower-level network that provided the input. The lower-level network then recalculates its output pattern with the new feedback information and sends the newly calculated pattern to the upper-level network. In the pure form of predictive processing, the lower-level network sends just the difference between the feedback pattern and its newly calculated pattern as an "error" pattern: using just the error data greatly reduces the amount of data that the network must send and thus is very efficient.[29] Whether biological networks in fact send an "error"

---

[28] Ned Block succinctly describes this logic of conditional probability in the visual system in particular:

> If h is the environmental hypothesis, e is the evidence from visual data and $p(h|e)$ is the probability of h given e, then $p(h|e)$ is proportional to $p(e|h) \times p(h)$. $p(e|h)$ is the 'likelihood' (of the visual data given the environmental hypothesis) and $p(h)$ is the prior probability of the environmental hypothesis.

The earlier model gives both $p(e|h)$ and $p(h)$ because, in theory, the network knows how the hypothesis-state works. Ned Block, "If perception is probabilistic, why does it not seem probabilistic?" *Philosophical Transactions of the Royal Society* B (2018), p. 7.

[29] Flores P. de Lange et al. very usefully point out that "predictive" processing does not entirely strip away the initial sensory data and rely on just error information. Nor does it attempt to simply predict what happens next but makes a probabilistic calculation about the nature (and cause) of the input that it has received so that it can anticipate what follows:

> A common point of confusion when describing predictive coding as a model of cortical function is the misconception that only prediction errors are encoded because the predictable part of the input is subtracted out by the prediction. While there are example systems where the predictable part of the signal is indeed removed (e.g., in the retina, to increase the dynamic range), predictive coding models of the cortex always contain separate neural populations representing both the current best guess (prediction) as well as the error associated with the guess (prediction error). Another common misconception is that predictive coding is a computational model of the cognitive process of prediction (or 'forecasting'). While it is certainly possible to model such a process within a predictive coding architecture (e.g., by changing the baseline activity of a specific prediction in anticipation of sensory input), predictive coding itself is a general theory of how the brain can efficiently encode

pattern after the first iteration of the data (recall that response times suggest that there can be about 100 steps in the cycling of data among layers of networks) seems less important than the idea that some approximation of probabilistic Bayesian inference is at work when the brain draws on memory to process the streams of sensory input.[30] In any case, if all turns out well and the state anticipated by the network turns out to be true, then the model is confirmed and strengthened, but if the input pattern in fact did not map onto the predicted pattern, then the network needs to adjust the probabilities associated with the "prior" in the model to get a closer match to the "world" as experienced.

Since the goal of predictive processing is not just to respond effectively to maintain homeostasis but to anticipate what will demand response, some researchers have introduced an additional term, "allostasis," to describe the task of preemptively dealing with change:

> [R]ather than merely responding to physiological perturbations in order to ensure the internal conditions of the body remain within homeostatic bounds, allostasis enables the organism to proactively prepare for such disturbances *before* they occur.[31]

Models that focus on allostasis propose a hierarchy of neuronal processing, with homeostasis serving as the base-level requirement and processes for allostasis then

---

information, and does not have the specific aim of explaining the cognitive process of prediction. In fact, most seminal models of predictive coding in the cortex do not include forward predictions: in these models, predictions are the result of an initial bottom-up analysis and are only formed after the first wave of feedforward activity. Such 'predictions' are thus hypotheses about the current sensory input, rather than forecasts of what is coming next

Floris P. de Lange, Micha Heilbron, and Peter Kok, "How Do Expectations Shape Perception?" *Trends in Cognitive Science* 22.9 (September 2018), pp. 772-73.

[30] Block, among others, has noted that the brain probably uses other techniques to do the calculations since the "kind of Bayesian computations that would have to be done are known to be computationally intractable." "If perception is probabilistic, why does it not seem probabilistic?" p. 8.

[31] Andrew W. Corcoran and Jakob Hohwy, "Allostasis, interoception, and the free energy principle: Feeling our way forward" in Manos Tsakiris and Helena De Preester, eds., *The interoceptive mind: From homeostasis to awareness*, (Oxford University Press, 2018), p. 2.

providing data to the homeostatic functions.[32] Although there are debates in the literature about this terminology and whether homeostasis can in fact encompass allostasis, I find the distinction useful since it stresses the fact that the world (including the body) constantly impinges on the brain in ways that it learns to anticipate in order to preserve its basic goal of maintaining homeostasis.

Once one models a hierarchy of neural networks to assess the state of the world and the body, then two additional concerns arise. The first is simply the *quality* of the data. Using the terminology of Bayesian logic, if $p(h|e)$ is the probability of h (the hypothesis about the world based on prior patterns) given e (the environmental data), the higher-level network's decision to adjust the probabilities associated with h (the hypothesis about the world) will depend on how reliable the new lower-level information is (its *precision*). As Karl Friston reflected, "most of the interesting bits of predictive coding are about getting the precision right: selecting newsworthy, uncertainty-resolving prediction errors."[33] The second point is that, given the significant functional differentiation among networks in different regions of the neocortex, while some networks use predictive processing, other networks need to be devoted to processing the *results* of predictions rather than making them.[34]

---

[32] As Corcoran and Hohwy explain,

> [T]he traditional notion of a homeostatic negative feedback loop is situated at the lowest level of the processing hierarchy, with its target setpoint (i.e., the expected physiological state) conditioned by information from higher (allostatic) circuits. These higher (or deeper) hierarchical levels are posited to model increasingly broader, domain-general representations of the present state of the body and its environment, as well as predictions about changes in those states.

Corcoran and Hohwy, "Allostasis, interoception, and the free energy principle," p. 11.

[33] Karl Friston, "Does predictive coding have a future?" *Nature Neuroscience* 21 (August 2018), p. 1019.

[34] Jakob Hohwy argues:

> When the explanatory focus is on real systems operating in the changeable and volatile world, PP [predictive processing] thus needs to move quite far beyond basic predictive coding. A PP system will be driven by its expected states, be characterised by a long-term, precision-sensitive perspective in both perceptual and active inference, and encompass model selection and complexity reduction in a rich hierarchical model. Different constellations of PP-processes speak to many different perceptual and cognitive phenomena, giving PP its extraordinary explanatory range.

IMPLICATIONS OF A PREDICTIVE MIND

Andy Clark and other researchers interested in the theoretical framework for neuroscience have come to stress a range of broader implications for how the brain operates if it in fact draws on predictive processing in its responses to the world. First, the line between perception and cognition blurs:

> In place of any sharp distinction between perception and various forms of cognition, PP [predictive processing model] thus posits variations in the mixture of top-down and bottom-up influence, and differences of temporal and spatial scale within the internal models that are structuring the predictions. Creatures thus endowed have a *structural grip* on their worlds: a grip that consists not in the symbolic encoding of quasi-linguistic 'concepts' but in the entangled mass of multiscale probabilistic expectation used to predict the incoming sensory signal.[35]

That is, perception always entails a rapid cognitive analysis of the input to assess the state of the world behind the input. As Clark notes, however, the cognitive analysis is in terms of hierarchies of structures of patterns. Very high levels of abstraction in semantic and episodic memory *can* be part of the processing, but they need not always be and mostly are not.

Clark, following Friston's model, goes even further in collapsing distinctions. The point of predictive processing is to allow the animal to correctly respond to what it encounters in its environment. The goal, that is, is planning for action. The predictive processing model tightly integrates this plan with the process of perception:

> Crucially, this implies that PP does not posit that brain activity includes only predictions and prediction errors; indeed, it may be that predictive coding occurs only at very low levels, with activity in higher levels given over to complexity reduction, and so forth.

Jakob Hohwy, "New directions in predictive processing" *Mind and Language* 35 (2020), p. 213

[35] Andy Clark, *Surfing Uncertainty: Prediction, Action, and the Embodied Mind* (Oxford: Oxford University Press, 2016), p. 107.

The emerging picture is one in which perception, cognition, and action are manifestations of a single adaptive regime geared to the reduction of organism-salient prediction error. Once-crisp boundaries between sensory and motor processing now dissolve, actions flow from percepts that predict sensory signals some of which entail actions that recruit new percepts. As we engage the world with our senses, percepts and action recipes now co-emerge, combining motor prescriptions with rolling efforts at knowing and understanding. Action, cognition and perception are thus continuously co-constructed, simultaneously rooted in cascading predictions that constitute, test, and maintain our grip on the world.[36]

The goals of predictive processing produce linked models of perceiving and "knowing" the world that underscore both these functions' creaturely origins and the fact that all that we perceive and know remain constructs assembled by the complex architecture and dynamics of the human brain:

> [W]hat we perceive is (when all is going well) the structured external world itself. But this is not the world 'as it is', where that implies the problematic notion (…) of a world represented independent of human concerns and human action repertoires. Rather, it is a world parsed according to our organism-specific needs and action repertoire. The world thus revealed may be populated with items such as hidden but tasty prey, poker hands, handwritten digits, and structured, meaningful sentences.
>
> Nor is there any sense in which the objects of perception are here being treated as anything like 'sense data' (Moore, 1913/1922), where they were conceived as proxies intervening *between* the perceiver and

---

[36] Clark, *Surfing Uncertainty*, p. 138. Clark returns to this theme later in his discussion: "One powerful strategy, which combines very neatly (or so I shall argue) with the image of the ever-active predictive brain, involves rethinking the classical sense-think-act cycle as a kind of mosaic: a mosaic in which each shard combines elements of (what might classically be thought of as) sensing and thinking with associated prescriptions for action." (p. 177)

the world. The internal representations at issue function *within* us and are not encountered *by* us.[37]

Andy Clark's first book, *Being There*, developed the argument that we should think of many of the ways we impinge upon the world as contributing to an "extended mind." In *Surfing Uncertainty*, Clark builds upon his earlier argument:

> In sum, our human-built worlds are not merely the arena in which we live, work, and play. They also structure the life-long statistical immersions that build and rebuild the generative models that inform each agent's repertoire for perception, action, and reason. By constructing a succession of designer environments, such as the human-built worlds of education, structured play, art, and science, we repeatedly restructure our own minds. These designer environments have slowly become tailored to creatures like us, and they 'know' us as well as we know them. As a species, we refine them again and again, generation by generation. It is this iterative re-structuring, and not sheer processing power, memory, mobility, or even the learning algorithms themselves, that completes the human mental mosaic.[38]

That is, we give external, perceptual form to structures we find conducive to human thriving. (Although Clark does not complicate this optimistic vision with all the other forms of human intention that perpetuate pain and domination, the dynamic of externalization remains the same regardless of content.) We in turn assimilate these external structures of patterns to shape the organization of the internal neuronal networks.

From Perceptrons to a Human-built World always Under Construction

The contemporary paradigms of predictive processing, with its systems of internal representations for the patterns the brain encounters and its model for

---

[37] Clark, *Surfing Uncertainty*, p. 195.

[38] Clark, *Surfing Uncertainty*, p. 279.

the mediation of perception through these representations, derive—in the end—from the successive elaborations of neural network models that began with the perceptron. Even though there are jumps, there also is an underlying continuity in the commitment to develop mathematical models to ever better represent the dynamics of neuronal systems in the brain. There is no magic here, but there is complexity and subtlety: the brain as modeled through predictive processing is designed to capture the textures and structures of the world and to order them not in some abstract, disembodied way but as a world as *humanly* experienced through the particular characteristics of individual bodies and of collective organization.

The theme of the profoundly embodied and situated character of human knowledge and experience as constructed within the neuronal systems of the brain underlies the explorations of specific systems in the chapters the follow.

Chapter Two

Connections: Introducing the Brain

Chapter one, on neurons and neural nets, examined the dynamics of what are, effectively, tightly coupled local networks. The focus of this chapter is the next level of structuring: that of the brain itself, a complex system of networks that are anatomically and functionally distinct yet participate in widely-dispersed patterns of activation that integrate separate functional units. The specific anatomical and functional organization of the human brain matters crucially in how it embodies meaning and selfhood as we engage the world. In this short chapter I introduce only the broadest outlines of the organization of the brain in order to provide the basic framework needed to explore the particular themes of memory and emotion that I believe should command the attention of humanists seeking to understand the microstructural elements of experience. The chapter begins with the basics—the parts of the brain—then moves to the functional organization of those parts.[1] The final section steps back to consider the rapidly developing framework of *network neuroscience* that explores the functional organization of the neuronal systems in terms of the paradigms of network analysis.

Introducing the Brain

What I present here is "the standard story" that is available in far greater detail in any neuroscience textbook. Much of the goal of this section is simply to

---

[1] While neuroscience has come to underscore the networked nature of regions within the neocortex in carrying out most cortical functions, the fine detail of exactly how neuroanatomy relates to functionally defined modules is still a work in progress. Nevertheless, it remains possible to associate particular functions with particular regions.

introduce the vocabulary needed to talk about the brain. (This section can serve as a handy reference for a quick look-up of terms that appear later in the book.)

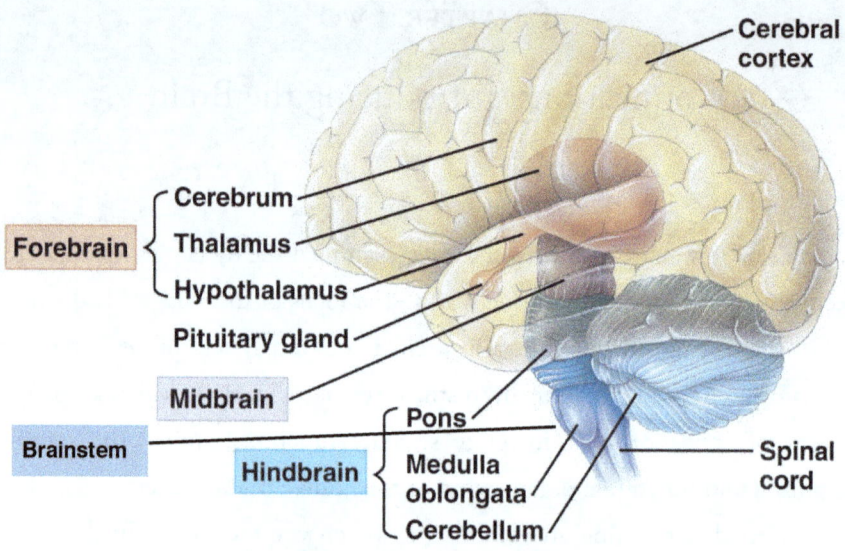

Figure 2.1: The Human Brain[2]

Developmental neuroscience traditionally divides the human brain is into three regions as it begins to form: the hindbrain, the midbrain, and the forebrain. The hindbrain—connected to the spinal cord—develops into the cerebellum, the pons, and the medulla oblongata. In the mature brain, biologists categorize the midbrain, the pons, and the medulla oblongata under the single term "brain stem." The cerebellum ("little brain") is a crucial and under-appreciated part of the brain: it is estimated that that while the human adult neocortex has around 25 billion neurons, the cerebellum has over 100 billion. However, since it primarily is involved in motor learning and control, which are outside the scope of this study, sadly, I too shall ignore the cerebellum. Throughout this book my focus will be on the brainstem and the forebrain.

I list the many structures in the brain here to suggest a crucial aspect to the story of the dissolution of the immediacy of experience. Namely, that the story is not simple. The list here of parts of the brain may seem long, yet these

---

[2] Image from "Nervous System," Slide 66 ("The Human Brain"), SliderBase, http://www.sliderbase.com/spitem-1447-9.html, accessed November 20, 2021.

are just the most basic structures. They all are important for the functioning of the hierarchy of neural networks in the cortex. In the neuroscience journal literature, for example, one constantly encounters observations like

> For language mechanisms, the cerebral cortex is the most important brain structure, although it is clear that its functionality crucially depends on subcortical input, in particular from the reticular formation, and functional interaction with basal ganglia, thalamus and limbic structures.[3]

The interconnectedness of the cortex and the subcortical and midbrain areas discussed below points to the distributed, mediated nature of the brain's neuronal systems, in which the more automatic "reflex" nuclei of the midbrain serve to guide the cortical modeling of the world and the body and to constantly modulate the interpretation of sensory and memory data by the networks centered in the cerebral cortex.

The Brainstem

We start with the brainstem and its evolutionarily old nuclei whose functions have been transformed through the long history of the development of the human brain. The midbrain, sitting atop the pons, contains a complex array of nuclei that serve vital roles in the maintenance of body functions and the relaying of somatic information to the cortical networks. These nuclei will reappear throughout the book because of their interaction with cortical systems, and especially with emotional networks. Chief among these midbrain nuclei are:

1. The superior and inferior colliculus. The superior colliculus is a complex, many-layered structure. It receives input from the optic nerves but also from both the primary and secondary visual areas of the cortex. Its output includes different parts of the thalamus (see below), including the lateral geniculate nucleus, which provides the input to the primary

---

[3] Friedemann Pulvermüller, "Neural reuse of action perception circuits for language, concepts and communication," *Progress in Neurobiology* 160 (2018), p. 3.

visual cortex.[4] The superior colliculus also has output (efferent connections) to the spinal cord. While in non-mammalian vertebrates, the optical tectum (the equivalent of the superior colliculus) is the primary visual region in the brain, in humans and other primates the superior colliculus, as a midbrain nucleus, functions in conjunction with the visual cortex in the visual system.

2. The ventral tegmental area (VTA) and the substantia nigra pars compacta are important in the dopamine "reward" system to be discussed in Chapter Four. The VTA receives input from a wide range of cortical and subcortical networks, but most notably the prefrontal cortex and the basal ganglia. The VTA has reciprocal connections (using dopamine) to the prefrontal cortex and the basal ganglia but also to many other cortical regions.

3. The reticular formation (including the locus coeruleus) contains over 100 nuclei responsible for transmitting information (especially about pain) to the forebrain and for maintaining various somatic systems. (The raphe nuclei, for example, regulate serotonin production.) Nuclei in the reticular formation are important in the control of attention, sleep, and consciousness.

4. The Periaqueductal Grey (PAG): a small sliver of neurons with complex connectivity and corresponding functions. It connects to the raphe nuclei, has many receptors for the important neuromodulators oxytocin and vasopressin, and appears to gate (i.e. increase, decrease, or entirely shut down) the sensation of pain generated in the cortex, among other functions. As will be discussed in Chapter Five, the PAG is active in many aspects of the emotional system and will reappear many times in this book.

As noted above, these midbrain nuclei have a long evolutionary history, but it is crucial to stress that they are part of a series of networks that integrate activations

---

[4] For a discussion of the role of the superior colliculus in the attentional control of eye movement, for example, see Richard Veale, Ziad M. Hafed and Masatoshi Yoshida, "How is visual salience computed in the brain? Insights from behaviour, neurobiology and modelling" *Philosophical Transactions of the Royal Society B* (2017), pp. 1-14.

by which cortical structures assess and send responses to the body. They generate first-order translations of a wide range of somatic homeostatic information into synaptic activation and neuromodulator production that the cortex uses to learn to link body states with other forms of experiential data.

The Forebrain

In fetal development, the forebrain initially divides into two regions: the diencephalon—which then forms the thalamus and hypothalamus—and the cerebrum. The cerebrum in turn consists of the basal ganglia and the cortex. And finally, the cortex includes the more familiar cerebral lobes of the neocortex and a few regions with specialized functions and architecture (the amygdala and the hippocampus). While the nuclei of the midbrain are complex in their roles in providing information to the brain about body conditions and, conversely, translating cortical activations into somatic signals, they do not participate in the functional systems of the cortex as immediately as the thalamus and the basal ganglia.

Forebrain
   1. Diencephalon
      A. Thalamus
         The thalamus seems to be the great gatekeeper of the brain. All sensory information passes through the thalamus (with initial processing within the thalamus) en route to the sensory cortices that then respond to the thalamic signals. The thalamus also plays crucial roles—that are not yet fully understood—in both sleep and consciousness.
      B. Hypothalamus
         The hypothalamus has a central role in regulating a wide variety of neuromodulators in the cortex as well as autonomic systems in the body (blood pressure, food-intake, body temperature, sweating, etc.). It is strongly connected to the amygdala in the cerebrum (see below), to the reticular formation in the midbrain, as well as to the central nervous system. While the hypothalamus crucially interacts

with the experience-dependent neural networks of the cortex, those networks use the hypothalamus as part of their signaling dynamics (both within the brain and to the body) in which the hypothalamus's roles are largely fixed.

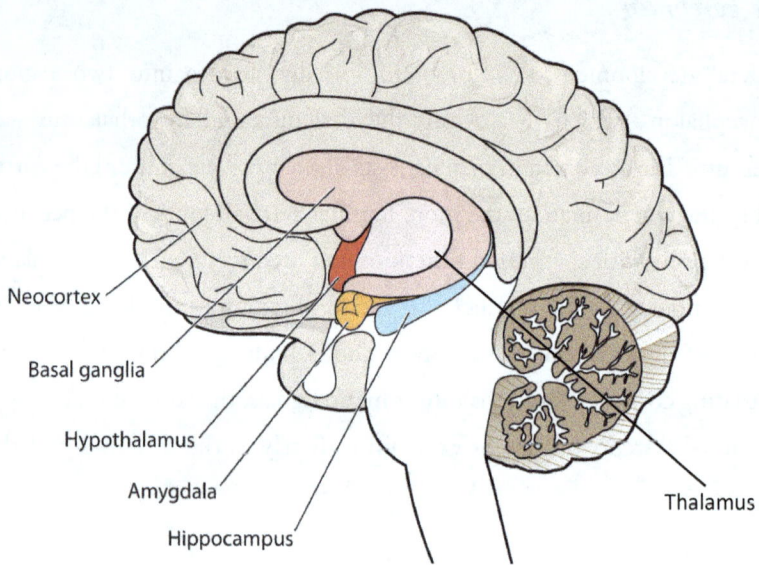

Figure 2.2: Medial View of the Brain[5]

2. Cerebrum

   A. Basal Ganglia

   As a matter of terminology, the basal ganglia, although cerebral structures, are referred to as "subcortical" since they are distinct from the cortex. They play a variety of roles, participating in at least five different circuits that connect to many parts of the cortex. While their role in the dopamine "reward" circuit is perhaps best known, they also appear to serve a central role in the selection of action (both motor action and more abstract forms of decisions) as well as in gating the activation of parietal working-memory association areas. One example of a motor system in which the basal ganglia

---

[5] https://www.123rf.com/photo_46940834_the-brain-in-cross-section-showing-the-basal-ganglia-hypothalamus-amygdala-and-hippocampus.html accessed November 20, 2021.

participate is the seemingly mundane but vitally important control of eye movement.[6]

B. Cortex

1. Hippocampus

   Although most of the neocortex is quite similar in its organization, the hippocampus, as part of the *allocortex*, has just four layers of cells (as opposed to six) and also has very distinct neuronal structures that are crucial to its role in rapidly recording and then gradually assimilating memories from often fleeting experiences. The hippocampus communicates directly to and is supported by the neocortical structures of the parahippocampal gyrus but also has connections to the amygdala and subcortical bodies. Chapter Six, on memory, will discuss the hippocampus in detail.

2. Amygdala

   Initially the amygdala was believed to be primarily associated with fear conditioning, but now is regarded as the cortical module for assessing emotional valence more generally.[7] Because it is important in emotion and in highlighting emotional valence information in memory formation, both Chapters Five and Six will discuss it at length.

3. The Neocortex

   The neocortex, the complexly folded sheet of grey matter that covers all the other components of the brain, provides the usual image we have of the brain. It consists of five (or six) regions called lobes, of which four are visible from the outside of the brain. I list the neocortex last because information gets to the it

---

[6] See the discussion of executive control networks in Michael I. Posner, et al. "Control Networks and Neuromodulators of Early Development" *Developmental Psychology* 48.3 (May 2012):827-35.

[7] For a recent study that expands the role of the amygdala, for example, see Melissa Malvaez, et al., "Distinct cortical–amygdala projections drive reward value encoding and retrieval," *Nature Reviews Neuroscience* 22 (May 2019):762-69.

through the filtering of all the subcortical structures described above.

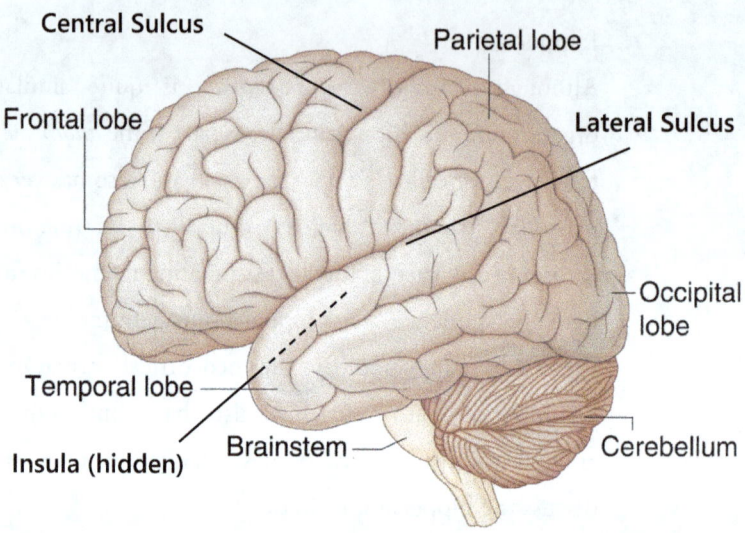

Figure 2.3: The Lateral View of the Neocortex[8]

1. The Occipital Lobe, in the back of the head, primarily processes visual information

2. The Parietal Lobe is above the occipital lobe and serves a variety of functions associated with working memory, memory, and the processing of associations between sensory data and memory about objects and events. In addition to these functions, the primary somatosensory cortex is on the ridge, the Postcentral Gyrus (Greek, "ring"), that is directly behind the Central Sulcus (Latin, "furrow"), the deep fold that divides the parietal lobe from the frontal lobe.

3. The Temporal Lobe, at the sides of the head, processes auditory information as well as handling long-term memory. The fold of the Lateral Sulcus divides the temporal lobe from the parietal and frontal lobes.

4. The Frontal Lobe, at the front of the neocortex, is the site of higher cognitive processes and also directs motor function. The primary motor cortex is on the

---

[8] Bernard J. Baars and Nicole M. Gage, *Cognition, Brain, and Consciousness: Introduction to Cognitive Neuroscience* second edition (New York: Academic Press, 2010), p. 145.

Precentral Gyrus, across the central sulcus from the primary somatosensory cortex in the parietal lobe. The prefrontal cortex, in the front part of the frontal lobe, has particularly important roles in short-term memory (and sustained attention) and executive control. The orbitofrontal cortex, in the lower part of the prefrontal cortex, is involved in the encoding and recall of emotionally valenced memories; it has connections to the VTA dopamine system, the amygdala, and the cortices around the hippocampus.

5. The Cingulate Cortex, sometimes considered to be a major part of a Limbic Lobe, sits on top of the *corpus collosum*, the bridge of white matter in the middle of the brain that provides the neuronal connection between the two cerebral hemispheres. The anterior cingulate cortex in particular provides an important form of working memory to process conflicting possibilities for action. The cingulate cortex is broadly connected to other networks to which it contributes high-order processing of incoming activations.

6. The Insula, increasingly considered the fifth lobe, is hidden out of view, on the other side of the temporal sulcus from the temporal lobe. The insula handles interoception (the information from the midbrain nuclei about body states). It maps information about pain in the body in the way that V1 maps retinal information, and it is important in emotional processing.

Finally, in addition to the names for parts of the brain, there are a few directional terms that also are a useful part of the vocabulary for talking about the brain and its various components. The dorsal part of a region is the *upper* part; the ventral is the *lower*. The rostral section is *in front*, while the caudal is *behind*. Figure 2.2 is a medial view of the brain, that is, seen as if one had split the brain apart where the right and left hemispheres divide. In contrast, Figure 2.3 is a lateral view, seen from the outside and from the side. Thus, medial sections of a brain region are those *closer to the center*, while the lateral sections are closer to the *outside and side*. Therefore, the *ventromedial* prefrontal cortex is lower and toward the center, in contrast to the *dorsolateral* prefrontal cortex, which is higher and to the outside.

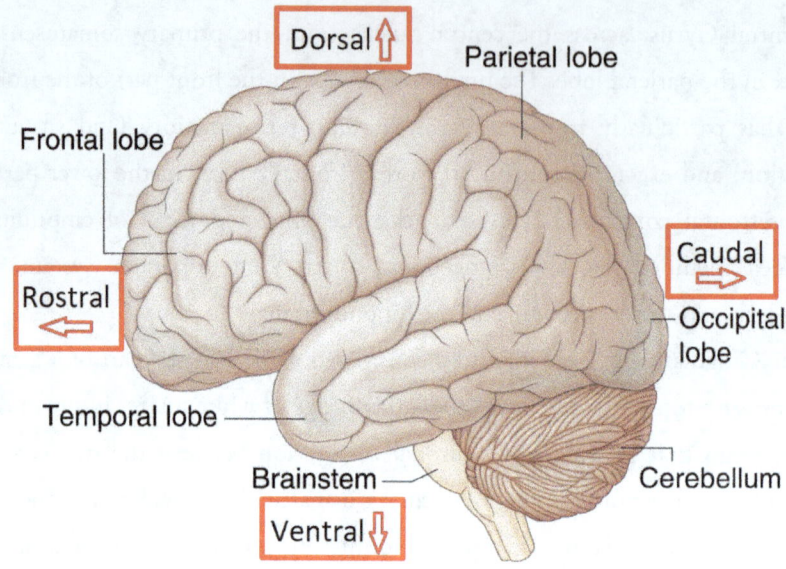

Figure 2.4: Names of Directions for the Brain[9]

Cerebral Cortex

When we discuss neural networks and how they structure themselves in response to experience, we almost always will be discussing neuronal systems in the cortex. Although our usual image of the "grey matter" of the cortex is of a rather massive lump of tissue, in fact the neocortex is only about 3-4 mm (0.12-0.16 inches) thick. As the cortex develops prenatally, it folds in a complex but well-structured manner. The folding—produced by the correct crimps appearing at the correct time as the brain grows—manages to get about 2,400 cm² (2.5 square feet) of cortex inside the average adult skull. The thin sheet of cells of the cerebral cortex has six layers. This seemingly minor fact turns out to have major implications for how the brain functions. The outermost Layer 1 has few neuron cells of its own and seems to serve primarily as a site for interconnections among neurons from other cortical regions. Layers 2 and 3 have both horizontal connections like Layer 1 and project below to other regions. Layer 4 is where most afferent (input) connections end. The innermost Layers 5 and 6 contain the major efferent (output) connections to subcortical regions and also participate in larger cortical

---

[9] Bernard J. Baars and Nicole M. Gage, *Cognition, Brain, and Consciousness: Introduction to Cognitive Neuroscience* second edition (New York: Academic Press, 2010), p. 145.

pathways.[10] Below Layer Six is the sheet of white matter, tracts of neurons that connect the neural network of the brain over long distances. "White matter" gets its name from the dense sheathing of fatty myelin around the axons of the neurons in the white matter that allows the neurons to carry spiking signals over long distances.

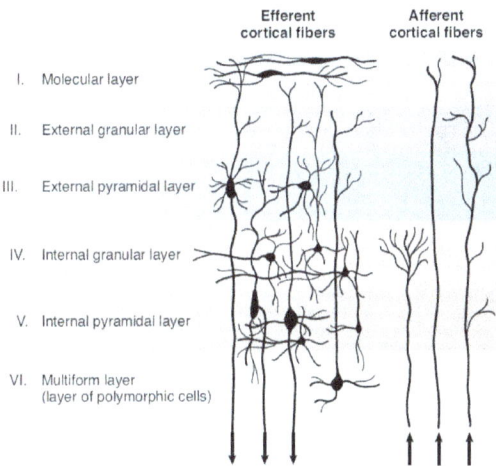

Figure 2.5: The 6-Layered Neocortex[11]

Historically, the regions of the cortex have been identified according to the system developed by Korbinian Brodmann (1868-1918), who divided the cortex into 11 regions with a total of 52 distinct areas, based on the microscopic details of structure and connectivity. Researchers today continue to identify regions of interest through their Brodmann area number. Brodmann, a neurologist, not unreasonably believed that there was a correlation of structure and function in the cortex, and for many years people labored to sort out the functional differentiation among the various Brodmann areas. Many of these

---

[10] Many books provide an overview of these structures. For an account that centers on the cognitive importance of developmental issues here, see Mark H. Johnson, *Developmental Cognitive Neuroscience: an Introduction* (Oxford and Cambridge, MA: Blackwell, 1997), pp. 24-39.

[11] The figure is Figure III-10-5, The 6-Layered Neocortex (p. 725) in *USMLE Step 1: Anatomy – Lecture Notes*, edited by James White and David Seiden (Kaplan Medical Books, 2018). © Simon & Schuster, 2018.

identifications are preserved in the contemporary understanding of the functional organization of the brain. For example, we have:

> Brodmann area 1, 2, and 3 in the postcentral gyrus of the parietal lobe:
>   identified as the somatosensory cortex.
> Brodmann area 4 in the precentral gyrus of the frontal lobe:
>   identified as the principal motor area
> Brodmann area 17 and 18 in the occipital lobe:
>   identified as the primary visual cortex.
> Brodmann area 41 and 42 in the temporal lobe:
>   identified as the auditory cortex
> Brodmann area 44 in the temporal lobe:
>   considered to be Broca's area for language in humans

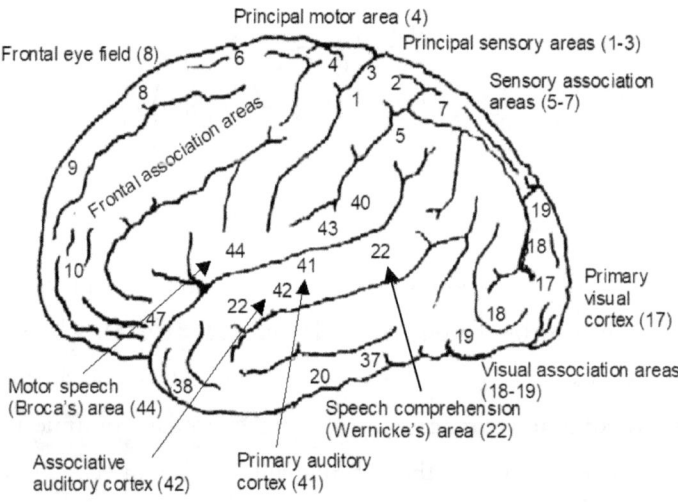

Figure 2.6: Functional Roles Identified for Brodmann Areas

While Brodmann's anatomical system was a major contribution to the study of the brain, our current understanding of the relationship between large-scale cortical structure and function draws on three sources that were not available to Brodmann: lesion studies, direct electrical probing of the brain and, more recently, various forms of brain imaging.[12] Over the years, neurologists accumulated a wealth of detailed information on the behavior of patients with brain injuries. These data allowed researchers to draw approximate functional mappings. If the loss of a section of the brain led to a loss of previous abilities—the ability to read, for example—then surely, the argument went, that brain area participated in the systems that made reading possible. There were doubts about this simple correspondence, and researchers were aware of cortical plasticity. Still, the account was plausible and reasonably well supported by the preponderance of the data. Furthermore, the work of Penfield and Rasmussen, reported in 1957, tended to confirm these findings. Their experiments used electrical current to directly stimulate brain tissue of conscious patients who reported their responses. They probed the regions on either side of the central sulcus and confirmed that the ridge in front of the sulcus was the primary motor cortex, and that behind the sulcus was the primary sensory cortex. Working their way from one end of these regions to the other, they determined what neurons caused what part of the body to either twitch (primary motor) or report sensation (primary sensory). They created the famous "homunculus," a drawing that illustrates the relative sizes of the neural real estate devoted to each part of the body.

Similar techniques explored other areas of the cortex where previous lesion studies identified functional mappings. The results confirmed the earlier data and held out the possibility of developing a detailed understanding of the organization of the cortex. As techniques improved and more studies accumulated, however, this elegantly simple picture began to unravel. Detailed analyses of lesions in a large population of patients gave inconsistent results. Neurologists developed staining techniques that allowed them to trace what neurons communicated to what regions. It turned out that the direct interconnectivity between regions was much broader than the simple picture of localized functions would allow. Moreover, studies in transcranial stimulation

---

[12] For a good perspective on the history of understanding brain function, see Edwin Clark and Kenneth Dewhurst, *An Illustrated History of Brain Function* 2nd edition (San Francisco: Norman, 1996).

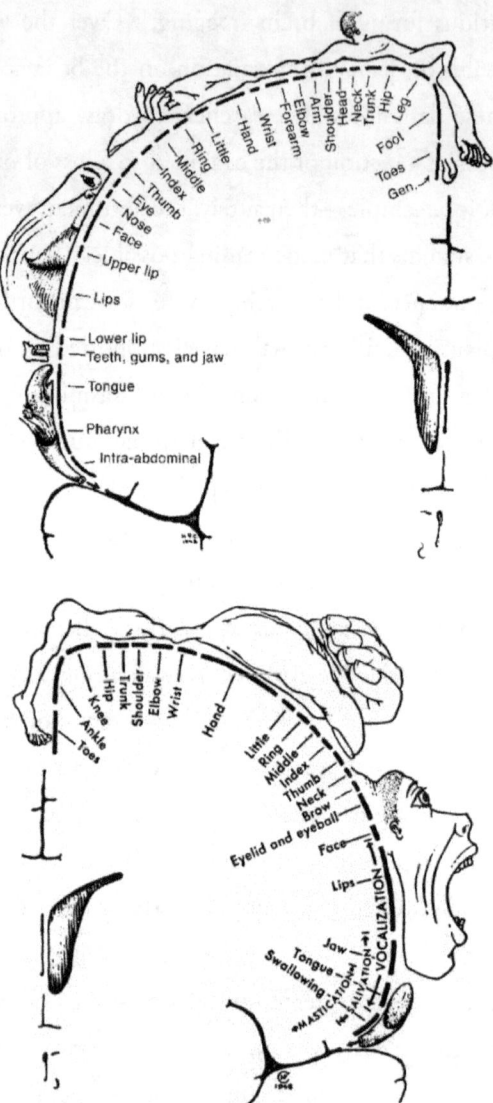

Figure 2.7a: Sensory Homunculus   Figure 2.7b: Motor Homunculus[13]

began to confirm the implications of the neuroanatomical data. The most recent development has been brain imaging through CAT scans, positron emission tomography (PET), and various forms of nuclear magnetic resonance imaging (MRI). These techniques are still works in progress, but they tend to repeat the disquieting results. When experimental subjects are given a task—reading,

---

[13] From Wilder Penfield and T. Rasmussen, *The Cerebral Cortex of Man: A Clinical Study of Localization of Function* (New York: Macmillan, 1957).

hearing, or thinking about saying a word, for example—the appropriate areas light up, but so do others. In the last paragraph of *An Illustrated History of Brain Function*, the authors conclude:

> It is becoming increasingly clear that the operation of the nervous system during the performance of a particular function depends on the pattern of excitation and inhibition in neuronal populations that are widely dispersed in different regions of the cerebral cortex. The belief that distinct functions are restricted to distinct cortical regions, as envisioned in the earlier part of this century, has become increasingly untenable. Recent work has emphasized the plasticity of the cortex, which is not simply a hard-wired array of neural circuits but a dynamic system that is modified or reorganized by preceding activity and feedback, and the importance of parallel distributed cortical networks with mutual reciprocal interconnections between different cortical areas.[14]

Non-invasive imaging techniques have continued to evolve, and two technologies have been of particular importance in changing our understanding of cerebral functional and anatomical connectivity over the past twenty years. The first is diffusion tensor imaging, which measures the direction that water diffuses in tissue.[15] This technology has allowed researchers for the first time to get clear

---

[14] Clark and Dewhurst, *An Illustrated History of Brain Function*, p. 170.

[15] Joan Stiles & Terry L. Jernigan give a wonderfully clear explanation of the technology in their classic article, "The Basics of Brain Development" *Neuropsychology Review* 20 (2010), p. 342:

> Diffusion imaging measures the diffusion of water molecules through the tissue. A common use of diffusion imaging involves fitting, for each voxel, a mathematical function called a tensor, that estimates proton diffusion (motion) along each of 3 orthogonal spatial axes. Tensors from voxels in the brain with high water content, such as in ventricles, exhibit high levels of proton diffusion that has no preferred direction; i.e., the diffusion is random, or *isotropic*. Diffusion in gray matter voxels is lower but also relatively isotropic. However, in voxels that contain fiber bundles, the diffusion is higher along the long axis of the fibers. This directionality of the diffusion is usually measured as an index of *anisotropy*, usually as fractional anisotropy (FA).

images of the white matter tracts that connect areas of the brain to one another over long distances.

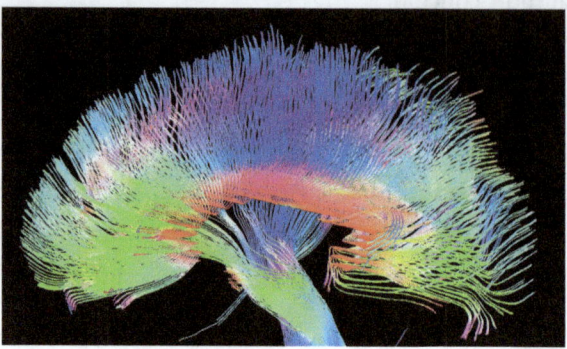

Figure 2.8: White Matter Tracts[16]

The second technology is an extension of functional magnetic resonance imaging (fMRI) called multivoxel pattern analysis (MVPA). A "voxel" in computer imaging is a point in space, the three-dimensional equivalent to a pixel, which is a point in two-dimensional space. In neuroimaging, however, it is a small unit of the fMRI image, covering an area larger than a cortical microcolumn so that one is not seeing the finest detail of neuronal behavior, which still requires invasive single-neuron probes. Still, the relatively fine-grained resolution of fMRI images allows researchers through the matrix mathematics of MVPA to compare images and discover differences in neural activity in specific regions and over large networks that are responding to the content being processed.[17]

Networks in the Cortex

When fMRI data made clear to researchers that neuronal processing was not local but widely distributed among the anatomical and functional regions of the brain, the next challenge was to understand the patterns in the distribution of activation for particular types of mental effort. Understanding white matter tracts was an important step in clarifying how activation spread across long distances.

---

[16] Tom Barrick, Chris Clark, SGHMS/ Science Photo Library / Getty Images Plus, https://www.sciencephoto.com/media/307225/view

[17] See, for example, Jeffrey R. Binder, et al., "Toward a brain-based componential semantic representation" *Cognitive Neuropsychology*, 33.3–4 (2016), 130–174.

Increasing mastery of the MVPA helped clarify what regions were especially active in particular tasks, given varying content.

As researchers began to probe the connectedness of networks, they came to realize the applicability of a realm of mathematics called graph theory. Graph theory provides a formalized account behind what is more popularly known as social network analysis (SNA) and has introduced a new descriptive vocabulary of networks, with terms like "hubs," "small world," and "rich club," as well as the basic terminology of nodes and edges. While complex systems analysis has helped researchers conceptualize and build models of recurrent networks at the local level, graph theory helps researchers think about the next level of integration that links more regional, more tightly coupled networks together to orchestrate higher-order processing in the brain.[18] One particular goal in network neuroscience has been to understand the so-called default mode network, the network of cortical regions that display coordinated activation when the brain is essentially idling, when a person is apparently not thinking of anything in particular. A recent article, for example, presents the DMN in network terms:

> Evidence emerging from such studies suggests that the default network comprises at least two separate networks with clear spatial distinctions along the posterior and anterior midline, which have often been described as hubs of convergence.... Each network comprises many spatially separate regions that fall throughout the association cortex. With a few exceptions, the nodes of one network lie side by side with those of another. One possibility is that this distributed and parallel organization might arise as a result of fractionation and specialization occurring at an early developmental stage; thus, a single, less-differentiated proto-organization becomes specialized during development by activity-dependent processes.[19]

---

[18] See Danielle S Bassett & Olaf Sporns, "Network neuroscience" *Nature Neuroscience* 20 (March 2017) 353-64.

[19] Randy L. Buckner and Lauren M. DiNicola, "The brain's default network: updated anatomy, physiology and evolving insights," *Nature Reviews Neuroscience* 20 (October 2019), p. 597

Researchers have identified a group of such networks that can be analyzed in graph-theoretic terms. Olaf Sporns, one of the most important figures in the field of network neuroscience has extended the application of network models throughout the connectome, the map of all the neural connections in the brain.

> An important aspect of the structural organization of the human connectome is the existence of a prominent 'rich club' [Figure 2.9b], defined as a set of highly connected and highly central nodes that are more densely interconnected than expected based on comparisons to degree-preserving null models. The cortical rich club included parts of the superior parietal cortex, the precuneus and posterior cingulate cortex, the anterior cingulate cortex, and the insula.... This finding suggests that the brain's rich club attracts and disseminates a large proportion of global communication, thus serving to integrate information across segregated communities and networks.[20]

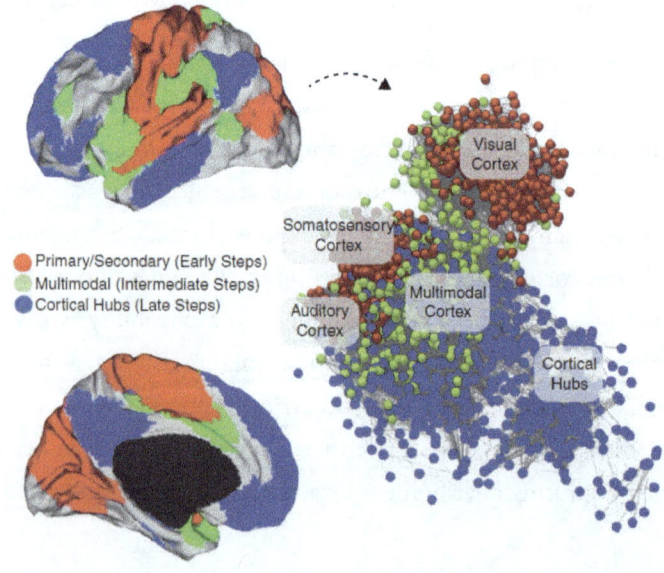

Figure 2.9 (a): Major Networks in the Brain

---

[20] Olaf Sporns, "Network attributes for segregation and integration in the human brain," *Current Opinion in Neurobiology* 23 (2013), p. 167. The image is Figure 3, p. 166.

The physical distribution of the major networks in the brain

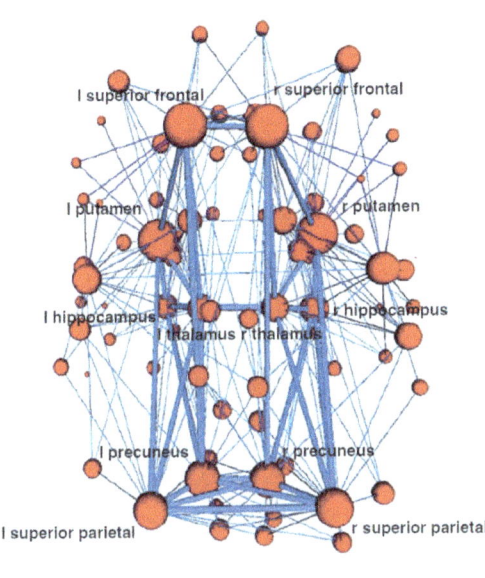

Figure 2.9 (b): Major Networks in the Brain[21]
The model of network connectivity among brain regions, including the core "rich club" connections.

This, then, is the contemporary view of the brain: a vast collection of nuclei and specialized regions that function as a totality by coordinated patterns of activation that move along networks with specific architectures. I return to these networks throughout this book, in the course of which the reader will become familiar with regions of the brain that play important roles in the story of emotion, memory, and selfhood.

---

[21] Olaf Sporns, "Network attributes for segregation and integration in the human brain," p. 166.

Chapter Three

Seeing the Light

Vision is our window to the world. The eyes report the real: open them, see a glass, reach for it, and the hand confirms that it is there. The glass is solid and cool; sip, and the mouth confirms that it is iced tea. Unfortunately, vision's direct access to the world that we experience every day is illusory (just as is the immediacy of touch and taste). While I *do* see, this chapter will suggest that it is better to state more cautiously that "I" "see": from the biological perspective, vision—even without the additional issues of visual consciousness, of an "I" seeing—is a complexly constructed phenomenon. The question, however, is what do humanists get out of delving into the complexity of the neuroscientific account of vision?

We know that seeing is not pure, that it is mediated. Indeed, visuality—a topic of much recent discussion in the humanities—refers specifically to the social and cultural mediations that construct what and how we see rather than to the intrinsic biological transformations of the data received from the retinas. This chapter will introduce a neuroscientific perspective in which the mediations that shape visuality—the ways in which external structures of social organization inform visual experience—are already a part of, and extension of, the ways in which memory and emotion, driven by creaturely biological commitments, shape visual processing. An understanding of the human visual system opens the way to grasping how the basic hermeneutics of experience begins with the integration of bodily, experiential (and statistical) meaning through the dynamics of neuronal networks. Exploring the visual system compels us to confront the inherent, unending polysemy of our most basic engagements with the world.

This chapter traces how the seemingly simple act of seeing draws on all the major cognitive and emotional systems of the brain and how explaining "what we see" requires an account of vision's integration into these other systems. When I see objects—that glass of iced tea, for example—I invoke memories: "glass" and "tea" as categories of "thing" as well as the tastes of tea and specific memories of my experiences with many sorts of tea. I assess what I remember and see compared with present desires, and I plan, not just abstractly, to move the muscles of the arm, hand and mouth. To explain "vision," I need to move beyond a simple account and to stress that the construction of what we know as seeing emerges from a synthesis of the activations of the most fundamental cortical and subcortical networks of the brain.

The simple story that still can serve as the basis upon which to build a fuller account of the visual system includes descriptions of the retina and the transmission of receptor activation information through the thalamus to the primary visual cortex (V1), information that then is transmitted to higher-level visual cortices, which synthesize the line-segment information provided by V1 into ever more complex shapes. As a final step, the visual system transmits information about visual objects to other parts of the brain which decide the significance of and proper responses to these visual objects. This simple version with which I begin is a "feed-forward account," based on the idea of the brain as a series of largely self-contained modules that receive data, process it, and pass it along to the next set of modules. The second thread of the chapter's account—in contrast—will address the ways in which the visual system is about much more than simply seeing. What we see in the world is shaped by attentional focusing and judgments about saliency that draw on memory, affective states, and larger goals. The visual cortex is in constant conversation with the memory, assessment, and control networks of the brain. This two-way conversation sharpens the visual data in its passage from the retina through the visual cortex to the higher cortical functions that must rely on what the visual cortex reports. This fuller account describes *embodied* vision—not seeing the world "as it is" but "seeing" cortical representations of the phenomenal realm shaped by the constraint of the human retina and by a complex system of neural networks trained through the history of interactions between human needs and experience in the world.

The Eye and the Retina

We begin with the human eye and the retina in particular.[1] The cornea, pupil, lens, etc. —the structures in the front of the eye that assure that photons reach the receptors at the back—deserve study in their own right, but for my purposes, the story begins where the photons activate receptors that then transmit neuronal activation to other cell types in the retina itself and then to the brain.[2]

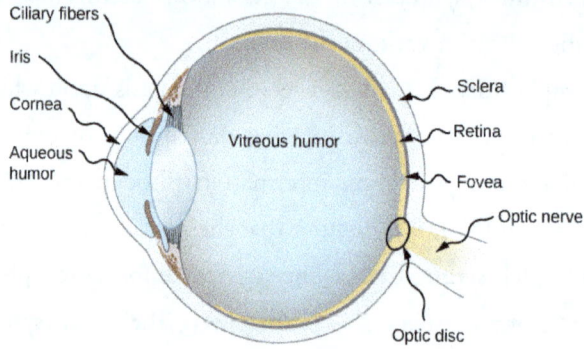

Figure 3.1: The Human Eye[3]

The retina consists of three layers of cells: ganglion cells, a middle layer (amacrine, bipolar, and horizontal cells), and a third layer: light-sensitive rods and cones. Rods, which are far more numerous, respond to dim light but quickly become saturated in bright daylight. Cones, in contrast, do not detect dim light well but are color sensitive and make it possible for us to see fine detail. Based on this difference in sensitivity, for example, there is a time early in the morning which in classical Chinese poetry is called *bian se* ("distinguishing color") when

---

[1] For an excellent comparison of the human retina with those of other species, see Tom Baden, Thomas Euler and Philipp Berens, "Understanding the retinal basis of vision across species," *Nature Reviews Neuroscience* 21 (January 2020):5-20.

[2] I base this account on David Hubel's *Eye, Brain, and Vision* (New York: Scientific American Library, 1995), pp. 36-57, and John E. Dowling *Creating Mind* (New York: Norton, 1998), pp. 101-22.

[3] The image is from Figure 2.6.1 of "The Eye" (2020, November 5). OpenStax CNX. https://phys.libretexts.org/@go/page/4496, accessed November 22, 2021.

the world is mostly shades of gray and the cones are just beginning to be able to register light and therefore color.

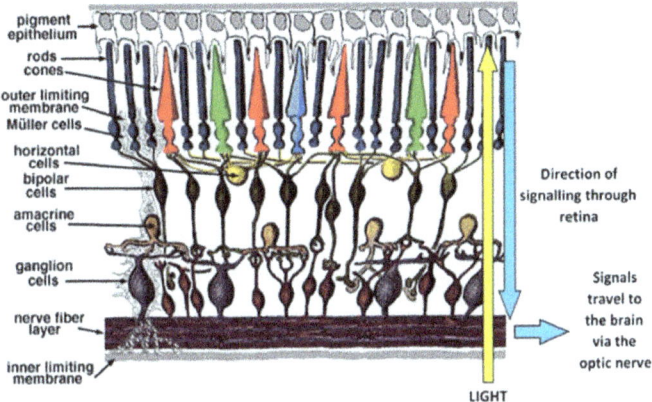

Figure 3.2: The Neuronal Structure of the Retina[4]

For humans and some other primates, there is a small spot at the back of the eye where the retina is all cones and no rods. This is the *fovea*, where vision is most acute. In the order of cell layers, rods and cones are *last*: two layers of cells (plus two layers of axons and dendrites) stand between the light and the detectors, an odd arrangement that is true everywhere except at the fovea, where the mediating layers are displaced and give the cones direct access. As a result of this displacement, the fovea appears as a tiny pit in the back of the eye, slightly above the center. Because the cones are densely packed and without interference at the fovea so that vision is the most acute there, humans and other primates scan the world by constantly shifting the fovea from place to place in the visual field and, if need be, move their heads to scan the scene more completely. Because of the density and accuracy of the information from the fovea, "a large percentage of the primary visual cortex (V1) is dedicated to processing its output."[5] Although

---

[4] Brittany J. Carr and William K. Stell, "The Science Behind Myopia", in Helga Kolb, Ralph Nelson, Eduardo Fernandez, and Bryan Jones, eds., *Webvision: The Organization of the Retina and Visual System* [Internet]. (Salt Lake City (UT): University of Utah Health Sciences Center; 1995- present). https://webvision.med.utah.edu/book/part-xvii-refractive-errors/the-science-behind-myopia-by-brittany-j-carr-and-william-k-stell/.

[5] Baden, et al., "Understanding the retinal basis of vision across species," p. 12

we constantly foveate—move the eye to allow the fovea to take in different parts of the visual field—this movement of the eye, called a saccade, is completely smoothed over in visual processing so that we are not consciously aware of it. Figure 3.4 shows a pattern in the unconscious foveation that traces the periphery of a picture of a face but returns over and over to the eyes.

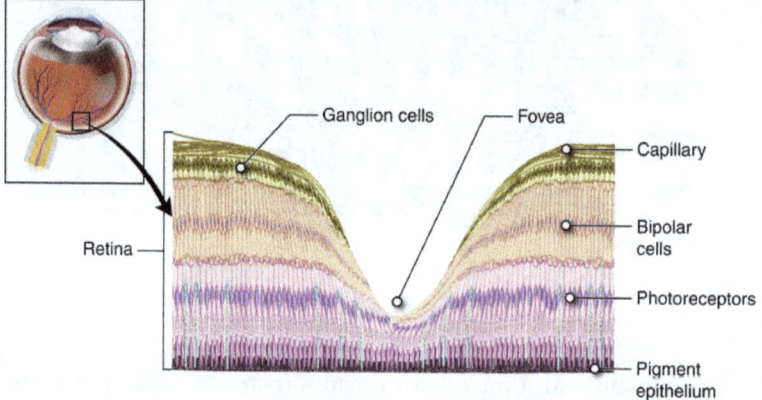

Figure 3.3: The Fovea[6]

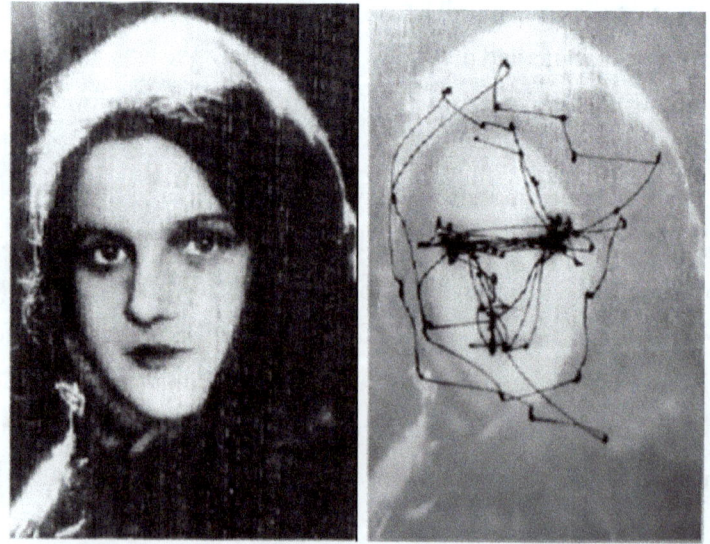

Figure 3.4: Using eye-tracking to record saccades[7]

---

[6] "Anatomy of the fovea - English labels" at AnatomyTOOL.org by Cenveo, license: Creative Commons Attribution.

[7] Alfred L. Yarbus, *Eye Movement and Vision*, trans. by Basil Haigh (New York: Plenum Press, 1967), Fig. 114, p 179.

The architecture of the retina—including the structure of the fovea—shapes important features of the human visual system, but the specific characteristics of our rods and cones also contribute to the particular nature of human vision. The response of the rods and cones to light varies by frequency, and the response curves in humans differ from those of other animals. The rods respond most strongly to green light (498 nanometers). Cone cells—which provide the data for color perception—have three subpopulations defined by their differing response curves: one group responds most strongly to red-yellow (564 nanometers), another to green (534 nm), and another to blue (420 nm). Because the tuning curve for the three types of cones overlap and thus, for example, the "green" cones respond a bit—just not as intensely—to the frequencies detected by the "blue" and the "red," all the colors of the visible spectrum can be represented by a combination of spiking rates of the three types of cones.

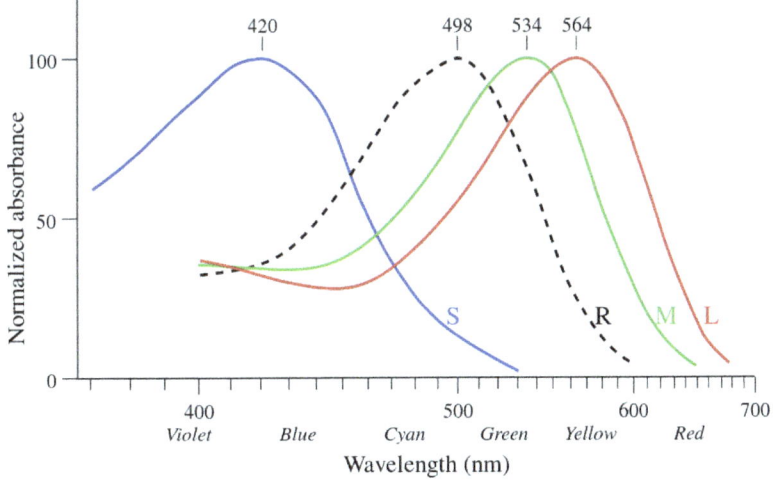

Figure 3.5: Photoreceptor Response Sensitivity[8]

---

[8] Vectorized version of the GFDL image Cone-response.png uploaded by User:Maxim Razin based on work by User:DrBob and User:Zeimusu., CC BY-SA 3.0 http://creativecommons.org/licenses/by-sa/3.0/ , via Wikimedia Commons, https://commons.wikimedia.org/wiki/File:Cone-response-en.svg (accessed November 22, 2021).

Both rods and cones transmit their activation to bipolar cells and horizontal cells. but the response patterns of the horizontal cells and bipolar cells reveal that a first layer of processing already has occurred. Horizontal cells are a class of interneuron, a laterally connected *inhibitory* neuron. They aggregate spiking information from the photoreceptors and, under certain conditions, can inhibit them (bright sun, for example). Bipolar cells, connected to the photoreceptors modulated by the horizontal cells, are either ON (*inhibited* from spiking by the photoreceptor when no light strikes them) or OFF (*firing* when they receive no light). They are connected to amacrine cells and to ganglion cells. The amacrine cells are another class of inhibitory interneuron connected to the bipolar cells and ganglion cells that modulate the responses of the ganglion cells in a variety of ways. Ganglion cells aggregate information from a group of photoreceptors as transmitted through the spiking of the bipolar cells and amacrine cells connected to them. Ganglion cells are (basically) of two types: on-center and off-center. An on-center ganglion cell spikes actively if light shines on the retinal receptors in the center of its "receptive field," and no light shines on the receptors around the edge. An off-center cell is just the opposite. Both types of cells, moreover, spike slowly in darkness and a bit more rapidly when both the center and surround cells are illuminated. The ganglion cells—through the mediation of the horizontal and amacrine cells—measure *relative* rather than absolute brightness, which turns out to be both a form of data compression and a way to assure decent acuity over a relatively wide range of lighting conditions. It is important to note that simple Hebbian training of such a system of receptors, bipolar cells, and ganglion cells with mutual inhibition supplied by horizontal and amacrine cells easily produces this pattern of activation. Amacrine cells also train ganglion cells with other types of responses. Connected in a network to a subpopulation of ganglion cells, for example, they train those ganglion cells through mutual inhibition to respond to *movement* in one direction or another. (That is, movement in one direction stimulates spiking; movement in the opposite inhibits it.) The on-center / off-center spiking patterns of the ganglion cells turn them into neurons that respond most strongly to edges where information changes.

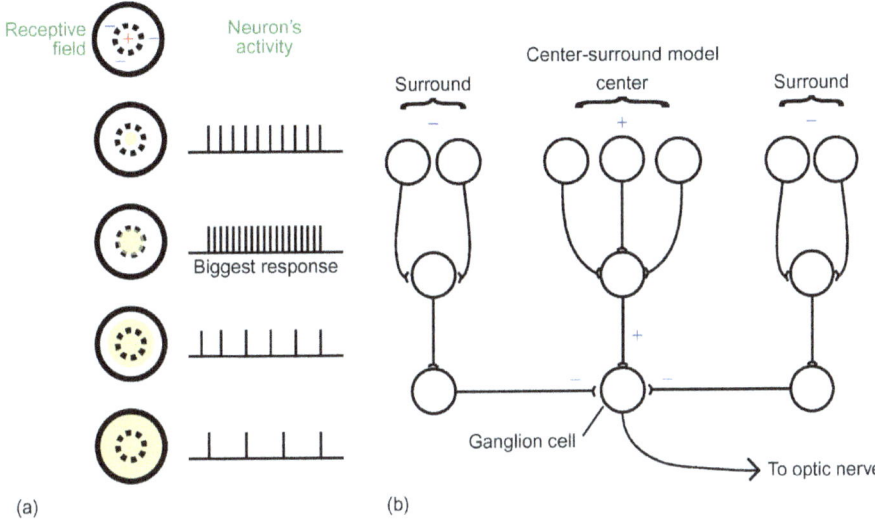

Figure 3.6: The Center-Surround Model[9]
(a) The response pattern of an on-center ganglion cell, (b) An inhibitory model to implement on-center ganglion cell response

Through these interactions, the retina performs a first level of data reduction that reorganizes the visual field:

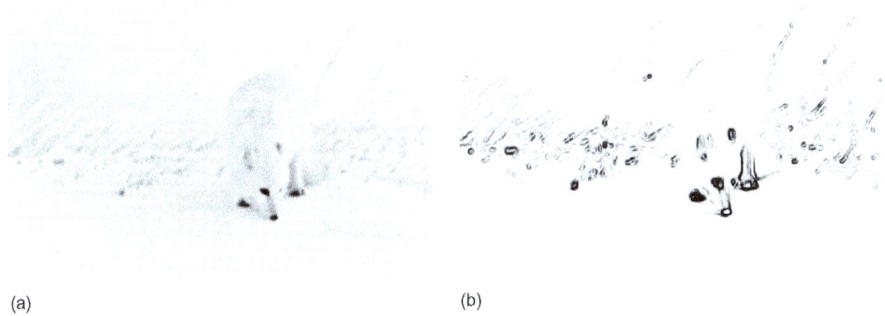

Figure 3.7: Image of Filtering to enhance edges in Ganglion Cells[10]

The axons of the ganglion cells form the optic nerve and transmit data that emphasize edges. Thus, the patterns of spiking transmitted through the optic nerve carry information about relative brightness, color, and movement that preserves the two-dimensional spatial mapping of the surface of the retina.

---

[9] Baars and Gage, *Cognition, Brain, and Consciousness*, p. 162, Figure 6.4.

[10] Baars and Gage, *Cognition, Brain, and Consciousness*, p. 162, Figure 6.5.

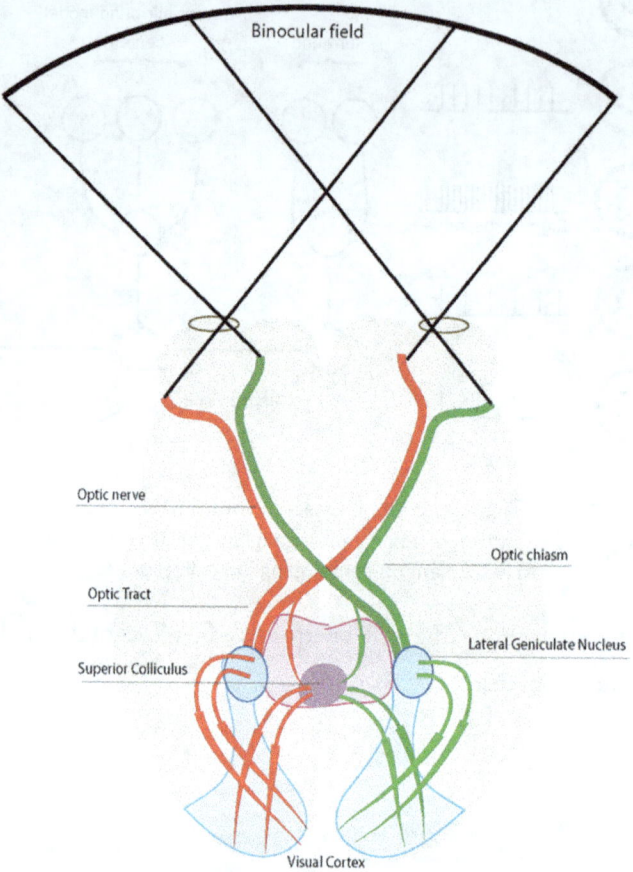

Figure 3.8: Binocular Visual Pathways[11]

The Lateral Geniculate Nucleus and the Primary Visual Cortex

The information from the ganglion cells next goes to a section of the thalamus called the lateral geniculate nucleus (LGN), which then feeds its responses to the primary visual cortex, V1. However, it is clear that a simple feedforward account of the LGN cannot explain its role in vision, since there are as many as ten times more neurons from V1 to the LGN than there are neurons from the retina and also ten times more neurons sending feedback from V1 than there are neurons from the LGN to V1.

---

[11] Figure from E. Herrera and C.A. Mason, "3.23 - The Evolution of Crossed and Uncrossed Retinal Pathways in Mammals," in *Evolution of Nervous Systems* Volume 3 (2007), Figure 2.

Scientists have long sought to understand the role of this strong feedback connectivity. It appears that the feedback to the LGN enhances spatial and temporal precision in the next cycle of activation it sends to V1. Spatial tuning is sharper and response times improve.[12] These may seem to be small matters, but the specific biological role of *attention* in the predictive–coding model is to improve the *reliability of information*, which this sharpening accomplishes. In addition to this role, the feedback to the LGN may more directly contribute to attentional modulation of the patterns of synaptic activation based on information transmitted from higher cortical levels outside the visual cortex, although this is not certain.[13]

While the lateral geniculate nucleus enhances the sharpness and the strength of the sensory signals sent via the optic nerve from the retinas, it still preserves (and clarifies) the basic patterns generated by the ensembles of retinal cells (i.e., unidirectional movement, dark center/bright surround, and bright center/dark surround). The primary visual cortex, then, has cells to effect a next level of aggregation. So-called "simple cells" in V1 respond to the angles of line segments formed through the clustering of information from individual retinal ganglion cells transmitted through the LGN. Some spike for horizontal bars; some prefer 72°; others react to 45° and so on. They use a system of "coarse coding": because the response of the cells is graded—falling off quickly for angles above and below the optimal angle for the particular cell, but still responding a little—linear combinations of a set of responses to segments at different angles can represent any possible orientation. As noted above, a similar logic is true for

---

[12] There had been early computational models for the interactions among the LGN, V1 and an additional area in the thalamus, the TRN. For example, John Bickle et al., "A Functional Hypothesis for LGN-V1-TRN Connectivities Suggested by Computer Simulation," *Journal of Computational Neuroscience*, 6 (1999):251-261. Strong biological evidence for these interactions had to wait until tools were developed to explore these systems. See J. Michaael Hasse and Farran Briggs, "Corticogeniculate feedback sharpens the temporal precision and spatial resolution of visual signals in the ferret," *PNAS* July 11, 2017.

[13] For a summary of the current understanding, see Farran Briggs, "Role of Feedback Connections in Central Visual Processing," *Annual Review of Vision Science* 6.18 (2020):1-22.

the response curves of the cones: output from cones tuned to only three colors are capable of representing all the colors of the visible spectrum.

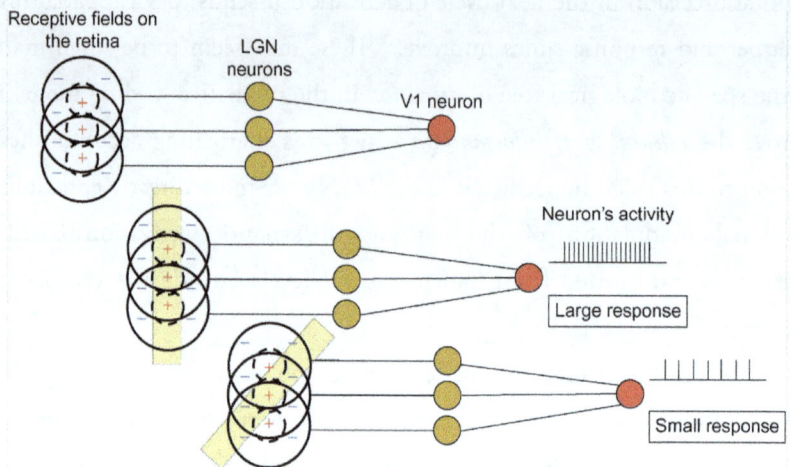

Figure 3.9: Orientation Selectivity in V1 [14]
The V1 neuron has been trained to respond most strongly to vertical line segments and weakly—or not at all—to line segments with different orientations.

The retinal response to light is built in to the genetically determined biophysics of the photoreceptors. In contrast, the organization of the simple cells emerges from a homogeneous network of neurons (with lateral inhibition) that is trained through simple Hebbian learning based on the synaptic patterns transmitted from the LGN. These networks are self-assembling feature detectors. However, since the retina also contains ganglion cells that respond to movement, there are corresponding movement detectors in V1. In addition, there are "complex cells" that combine movement detection with line orientation. Finally, among the complex cells, there seems to be a population that uses the simple-cell responses to develop a coarse coding for representative *two-dimensional* (rather than linear) features of the natural world as interpreted by the retina.

The organization of simple-cells (whose cell bodies are in layer 4Cα) and ever more complex complex-cells (whose cell bodies are in layer 4B) looks very much like the distribution of increasingly higher-order response patterns produced by introducing a wave of plasticity into the training of simple LISSOM

---

[14] Baars and Gage, *Cognition, Brain, and Consciousness*, p. 164, Figure 6.8.

systems discussed in Chapter 2. The inside-out maturation of the layers of the cortex—seemingly no more than a curious fact—may create precisely this logic of development. That is, at birth, when V1 begins to receive significant signals from the LGN, its neurons have a profusion of synaptic connections that begin to either strengthen or die and, through the interaction with the input from the LGN and from inhibitory interneurons, V1 forms its characteristic coarse-coded retinotopic neuronal network structure, first with the simple cells and then with the complex cells. The sensitive period for the training of the neurons in V1 is very early, before many of the higher-order cortical regions are developed (especially voluntary control of attention), and even before the maturation of V4, which provides feedback connections to V1 (see below).[15] Thus, the initial training of the neural networks in V1 comes largely from feedforward, lateral, and local feedback patterns of spiking.

One other feature of the primary visual cortex suggests the surprising behavior that can arise from self-assembling systems that confront streams of data from an implicitly structured environment. We have, of course, two eyes rather than one, and depth perception and stereoscopic vision are important for humans. The information sent from each retina, however, corresponds to its two-dimensional surface of receptors. The processing that makes stereopsis possible begins in the LGN, which has alternating rows of cells responding to each eye. In the primary visual cortex, this organization is more clearly developed: the arrays of simple-cell angle detectors are arranged in alternating rows for input from each eye. (Once more, it is simple competitive training that produces this result.) Moreover, there are complex-cells that build upon this organization and become disparity detectors. Some are "far-cells" that fire when a feature as it appears in one eye is further away than it is in the other. Some are "near cells," and some are tuned to precise convergence. To complete the picture, some of these complex cells fire only when there is movement as well.

Although in the initial feedforward account of V1, the simple cell assemblies respond just to the orientation of short line segments, feedback

---

[15] Zoe Kourtzi, et al., "Development of visually evoked cortical activity in infant macaque monkeys studied longitudinally with fMRI" *Magnetic Resonance Imaging* 24 (2006) 359–366.

connectivity complicates that model. From a theoretical perspective, according to the predictive coding framework, after the first bursts of the patterns of spikes that V1 transmits upstream, feedback should allow it to modify its spiking

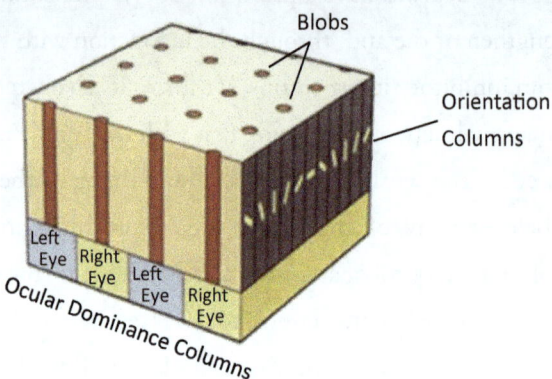

Figure 3.10: Ocular Dominance Columns[16]

patterns to reflect just the difference from the feedback activations that it receives. While there is not yet any experimental data to confirm this model, researchers have found other feedback effects that allow V1 to modulate its responses to LGN data. In particular, feedback from V4, combined with lateral connectivity within V1, helps V1 sharpen its representation of contour elements and to heighten the distinction between figure and ground in visual perception.[17] While in most models of V1, the receptive fields of simple cells are taken to be very narrow, spanning just a few retinal receptors, sharpening contours and figure-ground separation requires that groups of simple cells coordinate their responses, a coordination that is made possible by lateral connections among the V1 neurons.

---

[16] McGill: The Brain from Top to Bottom, Figure of hypercolumns, Copyleft https://thebrain.mcgill.ca/flash/a/a_02/a_02_cl/a_02_cl_vis/a_02_cl_vis.html (modified).

[17] Hualou Liang, et al., "Interactions between feedback and lateral connections in the primary visual cortex" *PNAS* 114.32 (August 8, 2017):8637-42. Also see Riggs, "Role of Feedback Connections in Central Visual Processing."

From the Primary Visual Cortex to the Visual System

The primary visual cortex, modulated by feedback from higher structures in the visual cortex, begins organizing the input stream into larger units of information. V1 reciprocally connects to V2, in which color, form, and motion appear to be processed separately, as well as to V3, V4, and the middle temporal area (MT, also referred to as V5). V2 then also reciprocally connects to V4. These higher levels create neural networks that respond to increasingly complicated patterns even as they dissociate sensory aspects into different streams.

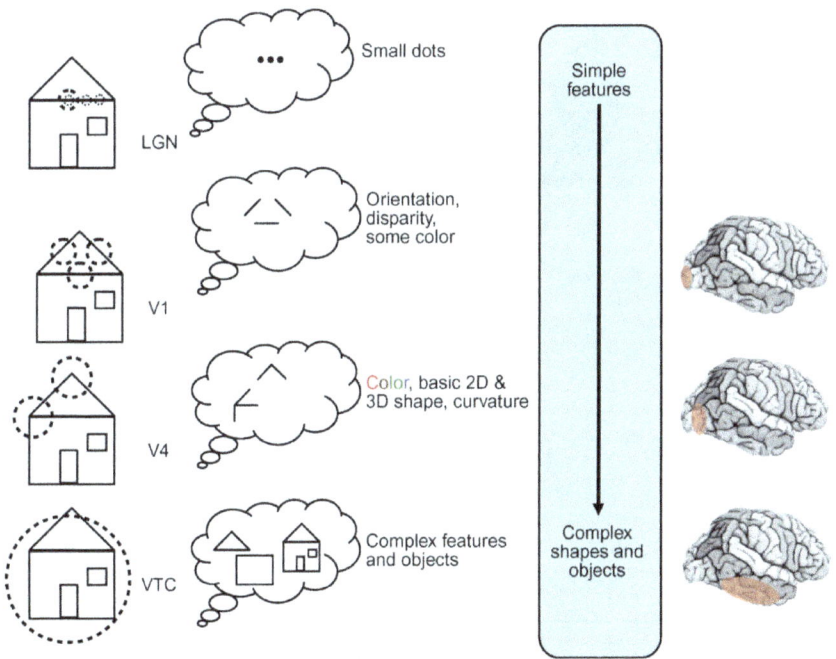

Figure 3.11: Increasing Complexity of Feature-Map Structural Units[18]

The role of feedback connections in V1 already suggests that the visual cortex is not an autonomous sensory processing module but deeply integrated into the brain in a complex dynamic system. Top-down recurrent activations modulate the feature-detectors at all levels in the visual cortex; feedback controls attention

---

[18] Baars and Gage, *Cognition, Brain, and Consciousness*, p. 166, Figure 6.10. VTC in the figure is the ventral temporal cortex. The colored area is the inferior temporal cortex (IT), which receives the output from MT.

within the visual field and seemingly underlies what we experience as visual consciousness both in active perception and in evoking visual imagery. Attention and consciousness introduce larger systems that underlie memory, recall, action planning, language, objects, and other such "upstream" phenomena. All of these systems participate in shaping visual processing.[19]

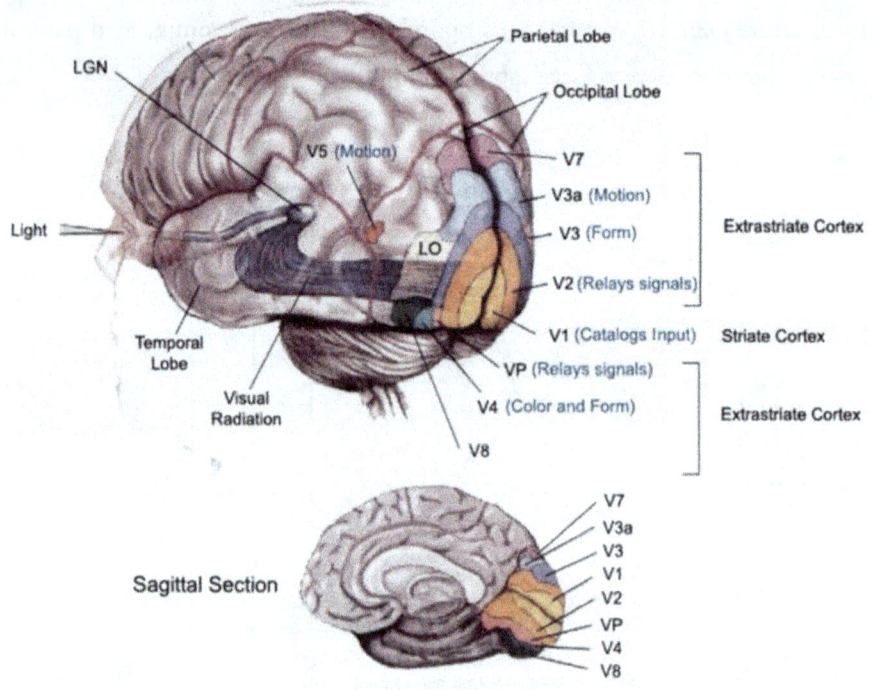

Figure 3.12: The Visual Cortices[20]

---

[19] Network neuroscience, which uses the growing accumulation of functional connectivity data to recast the interplay of regions across the cortex in terms of graph-theoretic networks, is profoundly changing thinking about how the brain works. In a review of the current state of the field, Danielle Bassett et al. observed:

> The sheer number of different kinds of network models used continues to grow, and their relationship with the kinds of relatively simple networks of interconnected neurons seen under Cajal's microscope becomes increasingly distant.

See Danielle S. Bassett, Perry Zurn and Joshua I. Gold, "On the nature and use of models in network neuroscience" *Nature Reviews Neuroscience* 19 (September 2018) 566-78. Also see Danielle S Bassett & Olaf Sporns, "Network neuroscience" *Nature Neuroscience* 20 (March 2017) 353-64.

[20] Nikos K. Logothetis, "Vision: A Window into Consciousness," *Scientific American* 281.5 (Nov. 1999), p. 72. © 1999 Terese Winslow LLC.

Attention

One example of these interactions between levels—despite the seeming simplicity of the task—will help us articulate the broad network of interactions that underlie what we know as vision. The control of tracking as the eyes follow a red ball bouncing across the floor requires a complex coordination of brain functions (see Figure 3.13).[21] A "ball" has properties that we know from experience. We know how it will bounce once we recruit both categorical (semantic) and experiential (episodic) memory. And we must stay on task: we must recruit attention.

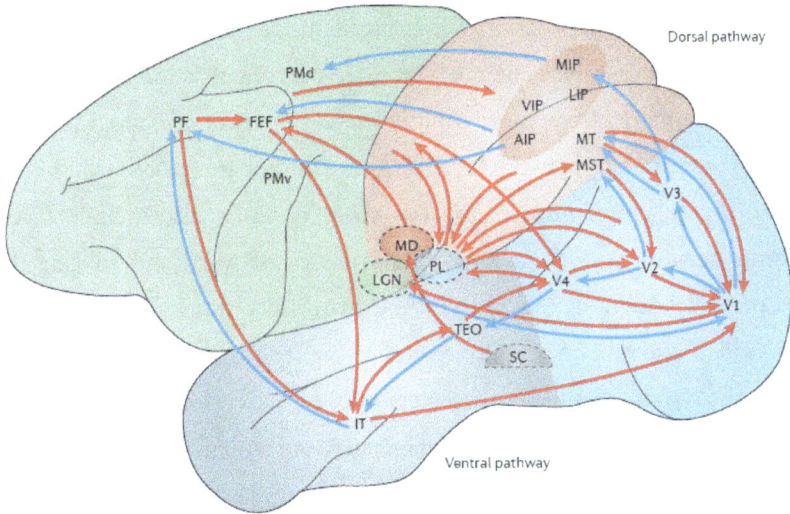

Figure 3.13: Reciprocal Connectivity to the Visual Cortex[22]

Consider even the most basic task in following the ball: the eye constantly moves to reposition the part of the visual field seen by the fovea, the small notch in the retina densely packed with cones. The decision about where to move the fovea next does not come from the visual cortex but from a system of neuronal

---

[21] Adapted from Mark H. Johnson, *Developmental Cognitive Neuroscience* (Cambridge, MA: Blackwell, 1997), p. 78. I rely on his discussion of the control of visual attention and eye movement.

[22] From Charles D. Gilbert and Wu Li, "Top-down Influences on Visual Processing" *Nature Reviews Neuroscience* 14 (May 2013), p. 351.

networks.[23] Two separate networks participate most immediately in the task of foveating: an attentional network that keeps the current task in mind and a salience network that assesses all the input in the visual field to identify objects and events that might warrant the interruption of the attentional focus. Because these two networks can conflict in their assessments, there is need for yet a third network to negotiate between attention and salience.

These three networks—one for control of sustained attention, one for assessment of the importance of the information flowing into the visual field, whether it currently is attended to or not, and a third system for the "control of control"—appear in slightly different forms in the current research literature.[24] A seminal three-part model proposed by Michael Posner, for example, includes a system for alerting, another for spatial orienting, and a third for executive control.[25] Contemporary versions of this three-part organization that have grown out of Posner's work offer a more complex series of networks that usually include a "dorsal attentional network" (DAN), a "frontoparietal control network" (FPCN) and a "cingulo-opercular network."[26] But they also include a ventral attention network as well as a separate salience network and suggest that a frontoparietal attention network (FPAN) is distinct from the FPCN. Figure 3.14—which attempts to visually represent the interactions of the networks—suggests the systemic problem: these networks are not entirely separate and share nodes, which means that activations of one network will necessarily activate those

---

[23] See the discussion in Richard Veale, Ziad M. Hafed and Masatoshi Yoshida, "How is visual salience computed in the brain? Insights from behaviour, neurobiology and modelling" *Philosophical Transactions of the Royal Society B* (2017)1-14.

[24] See the discussion in Evan M. Gordon, et al., "Three Distinct Sets of Connector Hubs Integrate Human Brain Function" *Cell Reports* 24.7 (August 2018):1687-95. The figure is from p. 7.

[25] Michael I. Posner and Steven E. Petersen, "The attention system of the human brain," *Annual Review of Neuroscience* 13 (1990):25–42.

[26] Although debates continue, the DAN seems to include the intraparietal sulcus (IPS) and the frontal eye fields), while the "frontoparietal control network" connects the dorsolateral prefrontal cortex and the temporoparietal junction. The "cingulo-opercular network" includes the anterior insula/frontal operculum and anterior cingulate/presupplementary cortex.

nodal regions shared with other networks. The *participation coefficients* (PC)—the degree to which nodes are parts of multiple networks—is especially revealing. Most of these networks, moreover, have a complementary relation to the default mode network (DMN), which is active when there is no particular focus for attention and grows inactive when the DAN becomes active.

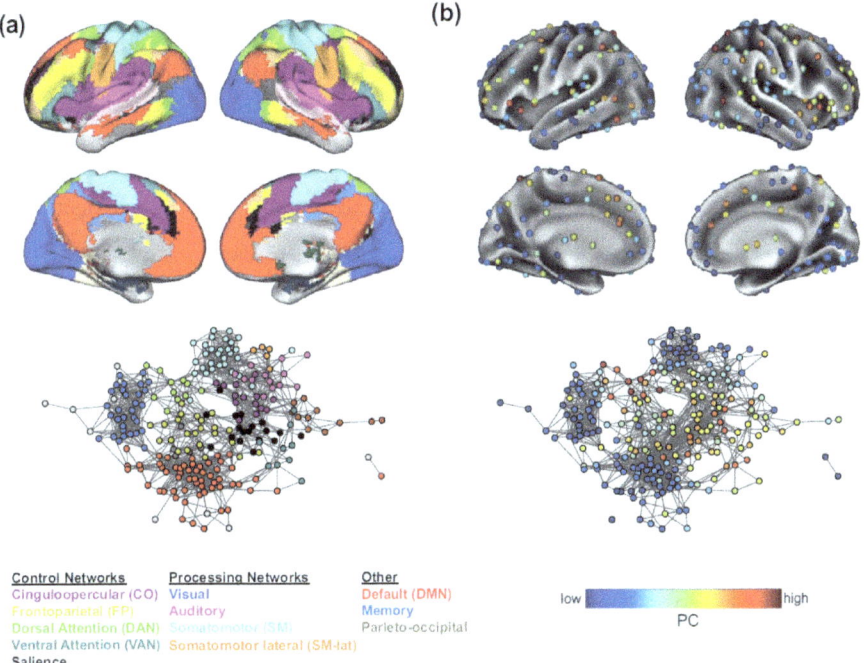

Figure 3.14: Neuronal Control Networks and Hubs[27]
(a) In the lower network diagram, the separate regions are represented as nodes, and the stronger the connection between nodes, the closer they are. Different colors represent different networks. (b) The networks are the same as in (a), but the color reflects the participation coefficient, the degree to which nodes participate in different networks.

Future research will tease out the dynamics of the attentional control that feeds back into the visual system to plan the next saccade and sharpen the input from LGN and V1 areas. In the meantime, the current research literature does provide a consistent list of input dimensions that shape attention and

---

[27] Caterina Gratton, Haoxin Sun, and Steven E. Petersen, "Control Networks and Hubs" *Psychophysiology* 55.3 (March 2018), DOI 10-1111/psyp.13032, p. 7, Figure 2.

salience.[28] First is object identity: data from the visual cortex reaches the inferotemporal and medial temporal cortical memory systems for objects and events. These in turn are connected to affective memory and affective assessment networks. Next, this information, with its affective component, needs to be weighed in working memory with the assessments—both long-range and moment-by-moment—of plans, goals, and means. The object-assessment data (1) could contribute to current plans maintained in working memory, which would require an update, or (2) the data could be such that it interrupts the inward-directed default mode network's activation and brings attentional resources into play, or (3) the new assessment may outweigh current plans and require a more significant reset, or (4) the data could simply be deemed a distraction.[29] In the first three cases, feedback to the early visual cortex next

---

[28] "Salience," however, turns out to be a tricky word because it has two quite different meanings. One describes the assessment of the bottom-up features of the visual field as reported by the retina (luminance contrast, color, motion, and orientation). The second involves the top-down assessment of the importance of the "objects" identified in the visual field that might merit an immediate shift of attention. It has been suggested that in early infancy, before connections to more rostral regions are well developed, the first version of salience based on features of retinal data serves as the alerting system, but as both memory and control capabilities rapidly develop, the second, experientially tuned version of salience becomes more important for disrupting and resetting the attentional control networks. See, for example, Dima Amso and Gaia Scerif, "The attentive brain: insights from developmental cognitive neuroscience" *Nature Reviews Neuroscience* 15 (October 2015):606-19.

[29] Much has been written about these core attentional tasks, and there is a vast bibliography. I have found several recent articles particularly useful. For cortical control network interactions, see, for example, Matthew L. Dixon, et al., "Heterogeneity within the frontoparietal control network and its relationship to the default and dorsal attention networks" *PNAS* DOI 10.1073/pnas.1715766115 (January 2018):E1598-1607; Taylor W. Webb, et al., "Cortical networks involved in visual awareness independent of visual attention" *PNAS* 113.48 (November 29, 2016):13923-28; Kristafor Farrant and Lucina Q. Uddin "Asymmetric development of dorsal and ventral attention networks in the human brain," *Developmental Cognitive Neuroscience* 12 (2015):165-74; C. Gratton, H. Sun, and S. E. Petersen "Control Networks and Hubs" *Psychophysiology* 55.3 (March 2018), DOI 10-1111/psyp.13032; Dima Amso and Gaia Scerif, "The attentive brain: insights from developmental cognitive neuroscience" *Nature Reviews Neuroscience* 15 (October 2015):606-19; and Radek Ptak "The Frontoparietal Attention Network of the Human Brain: Action, Saliency, and a Priority Map of the Environment" *The Neuroscientist* 18.5 (October 2012):502-15. Although I have largely focused on cortical networks, figure 3.13 also includes the thalamic clusters (the pulvinar, the medio-dorsal nucleus as well as the LGN) discussed

would increase the precision of the relevant network activations relaying the object information; in contrast, if the object is a distraction, the feedback inhibits that particular signal in the early visual system. All of these processes that spread activations across many interlinked networks are integral to the usual process of "seeing."

Finally, we need to recall that meaningful visual data have to be remembered. Thus, working memory and the visual association areas need to retain the activation patterns for the visual data and its contextual assessments long enough for the episodic memory system to process the new information.[30] The process of recording *into* episodic memory has its counterpart, the inverse process within the visual system: the recollection of visual information. This of course is very often a part of the assessment of ongoing visual data, but, equally crucially, recollection can occur "offline," in the absence of current visual input.[31] Such recollections both of events from episodic memory and of objects from "object memory" invoke many of the same networks and certainly go as far into the early visual cortex as V4, although with particularly intense imagining, activation may extend to V1.[32]

---

in Michael M. Halassa and Sabine Kastner, "Thalamic functions in distributed cognitive control," *Nature Neuroscience* 20 (December 2017) 1669-79.

[30] See, for example, Maya L. Rosen, et al., "The Role of Visual Association Cortex in Associative Memory Formation across Development" *Journal of Cognitive Neuroscience* 30.3 (March 2018):365-80.

[31] See, for example, the discussions in Charles E Connor & James J Knierim, "Integration of objects and space in perception and memory" *Nature Neuroscience* 20.11 (November 2017):1493-503, and in Carlo Sestieri, Maurizio Corbetta, Sara Spadone, Gian Luca Romani, and Gordon L. Shulman "Domain-general signals in the cingulo-opercular network for visuospatial attention and episodic memory" *Journal of Cognitive Neuroscience* 26.3 (March 2014):551-68.

[32] One surprising research result seems to be that the intensity of visual images is inversely correlated with the size of V1: the smaller the V1, the more intense the imagery. See Joel Pearson, "The human imagination: the cognitive neuroscience of visual mental imagery" *Nature Reviews Neuroscience* 20 (October 2019):624-34.

The Evolutionary Perspective

Over its long evolution, the human brain has developed an elaborate system of control networks that draw on all its resources to make sure that, at any given moment, we are looking at the right thing. What we see through this process of structuring visual input reflects particular details of the environment in which humans evolved as well as particular human requirements, both at the level of satisfying distinctive individual bodily needs and at the level of ensuring emergent behavioral patterns conducive to survival in the ever more complex human groupings that developed over the evolutionary time period.

Clear examples of these adaptations include the parahippocampal place area (PPA) that, in coordination with the prefrontal cortex, serves to process the details of complex environments, as well as the fusiform face area (FFA) that, as the name suggests, primarily serves to distinguish individual faces from one another—a need that grew as humans formed larger social groups. Such adaptations are pervasive and appear in small details of cellular behavior—as in the ganglion cells of the retina—as well as in aspects of white matter architecture, and in specialized functional units like the PPA and FFA that are integrated into the larger neuronal networks for vision.

Over evolutionary time, networks supporting vision have shaped themselves to reflect the human relationship to the encountered world. Researchers who do not start from computational neuroscience but instead explore the creaturely task that vision confronts as a biological system have arrived at conclusions that parallel the predictive approach and use the language of Bayesian probability:

> Because the information conveyed by our senses is both noisy and ambiguous, perception has often been conceptualized as a process of probabilistic inference in which the system decides on the most probable causes of our sensory inputs, based on the sensory data and prior expectations. Bayesian probability theory provides a principled way of making such inferences, dictating that agents should form and update their beliefs on the basis of not only the evidence provided by

the senses but also the prior probability of the various hypotheses about what is currently present in the world (i.e., expectations)[33]

That is, neurons themselves—as "leaky integrate and fire" devices—are noisy, and the visual field is complex and filled with only partially and fleetingly visible objects. The better the model of the experiential world that a creature can construct, the better it can interpret (and respond appropriately) to inherently incomplete data. The linking of these two facets of vision—internal noise and external complexity—drives much current thinking about how the brain handles perception. The contextualized, situational nature of visual experience informs the effort to see how the brain makes a multimodal model of the world to help guide visual processing.

> Two main concepts are central to current models of perception. The first is that to minimize errors, optimal perceptual decisions need to overcome uncertainty in both the external stimuli (for example, low-intensity sounds in the context of background noise) and the internal representations of those stimuli (for example, intrinsic variability in the firing of neurons encoding sensory information). The second is that, in naturalistic environments, stimuli rarely occur in isolation; they typically occur in the context of other stimuli, which concurrently or consecutively predict each other. For this reason, leveraging prior knowledge about the contextual relationships between stimuli can be advantageous for resolving sensory uncertainty and minimizing errors in perceptual decisions.[34]

---

[33] Floris P. de Lange, Micha Heilbron, and Peter Kok, "How Do Expectations Shape Perception?" *Trends in Cognitive Science* 22.9 (September 2018):772.

[34] Guillermo Horga and Anissa Abi-Dargham "An integrative framework for perceptual disturbances in psychosis" *Nature Reviews Neuroscience* 20 (December 2019):765. While the optimality touched upon by Horga and Abi-Darghan is important for computational neuroscience, Justin L. Gardner in a recent essay suggests that the mathematical bar of optimality here is too high. He agrees that for humans

> Neural mechanisms of perceptual decision-making have had the opportunity to be calibrated against the statistics of natural environments over evolutionary history

The visual system over evolutionary time developed solutions to satisfy changing human needs not through elegant approaches to optimization equations but in slow, messy adaptations that relied on the fundamental properties of neurons but repurposed some older parts of the system (like the superior colliculus) and added seemingly important new elements through the anything-but-teleological processes of genetic variation and natural selection.

"Seeing"

For creatures like us, "seeing" is more than capturing photons like a camera; research in the neuroscience of vision shows that "seeing" connects neural systems that synthesize activations from the retina in conversation with multisensory contextual information, memories, emotions, and the current state of the body. So far, I primarily have described the networks that participate in *controlling the focus* of visual information rather than what happens to that information as it is filtered through the control networks. We have gotten as far as the visual association areas of the intraparietal sulcus, V4, and V5 (MT), which seem to be the major nodes for processing the feedback activations from the higher-order control networks.

> and over the literally thousands of daily perceptual decisions we make as to where to look and how to interpret what we see. (p. 514)

However, reflecting on the stress on optimality in current mathematical models for perceptual decisions, Gardner suggests that human vision evolved through a more round-about, cumulative approach that produces solutions that are not perfect but good enough:

> One might worry that, without optimality theory to guide the search for neural implementations of perceptual behavior, we are left with simply a bag of tricks from which no principle can be discerned and no single model may apply. Likely there is some truth to this idea, as vision is the result of evolutionary pressures and constraints that have adapted it to solve species- and niche-specific problems and not as a perfect inference machine that re-represents the visual world.... Importantly, evolutionary and other pressures may only require satisficing of goals rather than optimizing. A solution that works under most circumstances may be good enough; visual perception can be fooled by illusions…, but the rarity of these special cases in which our visual system is in error is what makes them surprising and novel. (p. 519).

Justin L. Gardner, "Optimality and heuristics in perceptual neuroscience" *Nature Neuroscience* 22 (April 2019). "Satisficing" is the process of searching through available options until one that is minimally satisfactory (rather than optimal) is found.

As visual information is routed to higher-order networks, one standard model proposes that there is a dorsal pathway beyond the visual cortex that handles "where" information and a ventral pathway for "what" information.[35] In this initial model, the dorsal stream primarily feeds into planning for action, while the ventral stream contributes to cognitive functions. With improved tools and more extensive data, researchers in recent years have begun to move beyond

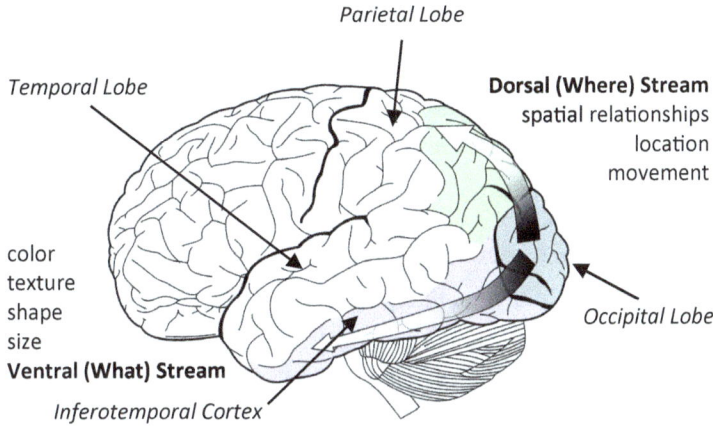

Figure 3.15: The Dorsal and Ventral Streams[36]

the cleanness of this functional separation. One argument is based both on data about functional connectivity and on more general issues of the organization of information:

> [Ours] is a world in which all things are spatial and most important things are objects. Even at the highest levels of perception and cognition, objects do not become disembodied abstractions characterized only by semantic labels. They remain real things whose detailed meanings are defined by their precise spatial structures and their spatiotemporal interactions with the rest of the world. Likewise, space itself is not experienced as an independent abstraction, but as a dimensionality that organizes and is organized by the ecologically

---

[35] The classic paper for this model is Melvyn A. Goodale and David Milner, "Separate visual pathways for perception and action," *Trends in Neuroscience* 15.1 (1992): 20–5.

[36] By Selket - I (Selket) made this from File:Gray728.svg, CC BY-SA 3.0, https://commons.wikimedia.org/w/index.php?curid=1679336, accessed 11/22/2022 (labels added)

important objects it contains. The ventral and dorsal pathways treat objects and space differently, but they cannot treat them separately.[37]

Current research on the two streams most certainly points to dorsal and ventral networks that serve distinct functions, but there is more cross-talk than the early model suggested. The ongoing project to construct a map of the human connectome—a map of all the interconnected networks of the brain—preserves the basic distinction between flows of information into the ventral memory systems and that into the dorsal motor areas. Still, recent work offers a more fine-grained model of visual information pathways that suggests some specific evolutionary adaptations in humans that support the complexity of our internal representational systems.

> There are several reasons why additional processing pathways might be expected in humans. First, the human brain, both in terms of its overall volume and the size of visual cortex, is much bigger than the primate brain, suggesting additional functions that might require additional or more specialised visual processing capabilities. Second, previous work suggests the existence of a large white matter fibre bundle, the inferior fronto-occipital fasciculus connecting human ventral occipital and inferior frontal cortex, that appears absent in non-human primates (…). Third, recent theoretical work proposed the existence of three major visual processing pathways in humans. For instance…, it has been proposed that the human cortical visual system comprises a dorsal occipitoparietal stream, a lateral occipito-temporal stream and a ventral occipitotemporal stream (…). In this model, the additional lateral stream, consisting of areas that are classically assigned to the dorsal stream, incorporates different aspects of vision, action and language.[38]

---

[37] Charles E Connor and James J Knierim, "Integration of objects and space in perception and memory" *Nature Neuroscience* 20.11 (November 2017):1493-503.

[38] Koen V. Haak and Christian F. Beckmann, "Objective analysis of the topological organization of the human cortical visual connectome suggests three visual pathways" *Cortex* 98 (2018):74.

I return to the lateral occipitotemporal stream for language in Chapter Six, and here want to focus instead on the ventral stream that integrates the visual data into larger structures of meaning that are most directly relevant to accounts of experience in the humanities. The ventral stream connects the visual perceptual input not just with the basic categories of what an "object" is but also with the various accumulated memories in which that object plays a role. We do not perceive objects "by themselves" but as elements of lived experience enmeshed in networks of meaning.

In particular, focusing on the connectivity of the anterior inferior temporal cortex (see Figure 3.13) provides a clear illustration of the ways in which the visual networks participate in—are a part of—the central systems for the construction of integrated models for meaning. The ventral visual pathway in humans is bigger than V1 and the extrastriate visual cortex combined, and its tiered organization clearly extends the logic of the visual cortex, in which higher levels serve to aggregate the components of the visual field that are structured in the lower levels. V1 develops a retinotopic map of short, angled bars, while V4 constructs a map of far more complex naturally occurring shapes. The general approach to visual processing in the inferior temporal cortex (IT) is that this same process of aggregation continues as one moves from the back forward (posterior [PIT] to anterior [AIT]).[39] The anterior region of the inferotemporal cortex (aIT) has the most highly articulated visual representational structures, which it constructs from the data passed to it from earlier in the stream. However, vision serves creaturely goals, and it is in the anterior IT with its specific categorization of the visual data that the visual system is integrated with the other networks that give the visual data its meaning. Concerning color information, for example, Bevil R. Conway notes that

---

[39] Bevil R. Conway observes:
> One idea suggested by the model is that each stage along IT has a computational objective such as detecting members of a category (PIT) or recognizing members within a category (AIT), and these operations involve tuning to the stimulus dimensions that capture the most variance for those problems.

Bevil R. Conway, "The Organization and Operation of Inferior Temporal Cortex," *Annual Review of Vision Science* 4 (2018), p. 394.

> [The] anterior IT is richly connected with subcortical structures, including the striatum and amygdala, and with medial temporal cortex (Kravitz et al. 2013), including the hippocampus—brain structures that are likely called upon to mediate the memory and rewarding aspects of color.[40]

The rich connectivity of the anterior IT (aIT) and its role in integrating visual data into larger representational and planning structures is a constant theme in the neuroscientific accounts. First, there is a group of crucial subcortical pathways.[41] In addition, given the importance of predictive coding in vision, feedback connections from other *cortical* networks also play a crucial role in the final stages of visual processing in the anterior IT.[42] These networks structure the

---

[40] Conway, "The Organization and Operation of Inferior Temporal Cortex," p. 392.

[41] Dwight Kravitz, for example, discusses several clusters of subcortical connections:

> There are three major output pathways to subcortical structures, all of which are critical in forming associations between visual stimuli and non-visual information. First, the unidirectional occipitotemporo-neostriatal pathway arises from every subregion of the occipitotemporal network except V1 ... and supports the formation of links between stimuli and responses. Second, the occipitotemporo-ventral striatum pathway arises in aIT and supports the association with and processing of stimulus valence. Third, the projections from the occipitotemporal network to the amygdala arise primarily in aIT, but projections from the amygdala target almost every subregion of the network (occipitotemporo-amygdaloid pathway)....

Dwight J. Kravitz, et al., "The ventral visual pathway: an expanded neural framework for the processing of object quality" *Trends in Cognitive Science* 17.1 (January 2013), p. 37.

[42] Kravitz, et al. describe the array of *cortical* connections to the inferotemporal lobe:

> In contrast to the subcortical pathways, the cortical outputs are bidirectional and originate in aIT, although from different subregions within it. First, the occipitotemporo-medial temporal pathway arises from every region in aIT as well as the TEpv [posterior ventral TE]... and supports the formation of *long-term cognitive visual memories*. Second, the occipitotemporo-orbitofrontal pathway arises primarily from the STSv/f, TEav [anterior ventral TE], and area TGv granular and supports the *association of visual stimuli with reward*. Third, the occipitotemporo-ventrolateral prefrontal pathway arises primarily from the STSv/f with only a minor projection from TEad and supports *object-based working memory*.... The projections from aIT into the rhinal cortices carry visual information used in the encoding of *long-term memory of object quality*, explaining the numerous findings indicating complex visual selectivity in these regions. [Emphases added.]

meaning of the visual realm and of what we see moment by moment. This embodied visual "perception" thus is neither immediate nor purely visual.

Retrospective and Prospective

This chapter began with questions about the cortical events we call "seeing" and about the neuronally conceived content of sensory experience. The account at first stressed the self-organizing character of the visual pathway from the retina to the thalamus and the ascending areas of the visual cortex. But recurrent networks pervade the system from the top to the bottom, so the story extended to the mutually structuring interactions among perception, attention, and memory. As I discuss in Chapter Six, memory is continuous with the upper reaches of perceptual processing, and we have seen from the attentional system that what matters at the higher levels shapes what is salient at the lower level of cognitive maps.

The cortical substrate for sensory experience is an ordered web of rapidly ramifying activations where all the levels of featural abstraction and integration constantly talk to one another simultaneously. This is a world of shades and echoes and drifting focus: this is in fact the world we know from our own experience. It is richly interconnected but correspondingly highly mediated. Even the most direct encounter follows an inner trajectory shaped by complexly sedimented traces of past experience. The next question we must confront is what controls the logic of sedimentation. There are internal biases to sensory experience—neuroscientists refer to them as "biologically relevant" input—that must be at work from birth if the child is to survive. We know them more familiarly as emotions and the affective system. The next chapter explores the affective system, the biological biases that shape the emergent properties of that system we call our selves.

"The ventral visual pathway: an expanded neural framework for the processing of object quality," p. 39.

CHAPTER FOUR

Affect: The Feeling of Being

Emotion binds us to the world, not necessarily to the here-and-now but to all that we have experienced. Emotion shapes the self, mediates how we know the world, and prepares us to act in it. The neuroscientific understanding of emotion—like that of vision—has been profoundly transformed by the emergence of network neuroscience, predictive models, and the new forms of imaging analysis that have allowed researchers to see the whole-brain neuronal patterns of response that support these developing paradigms. Contemporary explorations of the neuroscience of affect reveal that emotional systems touch widespread networks throughout the brain. Affect pervades the processes that underlie our engagement with the world, and the breadth of affective network interactions largely erases any clear demarcation between cognitive and emotional systems. A recent survey of the role of emotion in cognition concludes that

> [E]motional information constitutes a feature that is built into our representations of the external world in a manner similar to such "objective" features as shape and color, and primary sensory representations are rapidly modulated by emotional learning.... Like other mammals, we are enmeshed in the world such that emotional outcomes are the central organizing feature of perception, attention, learning and memory, guiding awareness, and informing action.[1]

---

[1] Rebecca M. Todd, Vladimir Miskovic, Junichi Chikazoe, and Adam K. Anderson, "Emotional Objectivity: Neural Representations of Emotions and Their Interaction with Cognition" *Annual Review of Psychology* 71 (2020), p. 43.

While there are many accounts of emotion in neuroscience, the story I tell in this chapter follows synthetic models that combine two major positions about the basic nature of emotion, both of which have grown increasingly nuanced over the past decade in order to account for new data but which remain incomplete on their own.[2]

The first of these positions proposes that emotions are combinations of discrete basic emotions, each of which has a distinct biological basis that distinguishes it from the others.[3] The second position—usually described as constructionist—rejects the idea of such discrete basic emotions. Its proponents argue instead that there is an emotional system for assessing key qualities of *salience* and *valence* in information coming from both inner and outer sources. This general emotional assessment system draws on processes of cognitive evaluation to generate the specific features of a particular emotional response.[4] In this chapter I present a third approach that appears in contemporary work in affective neuroscience. It stresses the hierarchical networks that underlie what we know as emotional experience and that organize sub-networks spanning the entire brain. This model thus accommodates both discrete substructures underlying emotion as well as the crucial role for learned, socially informed categories that articulate the network at the highest levels that are accessible to consciousness.[5]

---

[2] One excellent, accessible and critical account of the development of theories about emotion in biology is the third chapter, "Universalism: The Life Sciences" in Jan Plamper, *The History of Emotions*, translated by Keith Tribe (Oxford: Oxford University Press, 2015), pp. 147-250.

[3] See Paul Ekman and Daniel Cordaro's essay "What is Meant by Calling Emotions Basic" in the volume *Emotion Review* 7.4 (October 2015):364-70. Ekman has been a major proponent of "basic emotions," and this essay offers a refined version of the argument.

[4] Lisa Feldman Barrett presents a summary of her case for the constructed nature of emotions in "The theory of constructed emotion: an active inference account of interoception and categorization" in *Social Cognitive and Affective Neuroscience* 2017:1-23.

[5] Jaak Panksepp's seminal study is *Affective Neuroscience: The Foundations of Human and Animal Emotions* (Oxford: Oxford University Press, 1998). However, in a debate among Panksepp and Mark Solm on one side and the cognitive neuroscientists Richard D. Lane and Ryan Smith on the other, Lane and Smith seem to have learned Panksepp's lessons and broadened the framework to accommodate the paradigms of network neuroscience. See

The basic conceptual framework for the chapter builds upon two related ideas. The first is the observation that humans are altricial animals that require a long period of nurturing after birth. At birth, the cortex knows little about the body and even less about the external world and must rapidly construct initial maps correlating the patterns provided by the senses with those provided by the body. The particular challenge is to identify the patterns and correlations that matter most. This is the role of the brainstem and midbrain nuclei that provide crucial interoceptive information that mark current experience as important and worthy of particular attention in shaping the slowly emerging sensory and executive networks in the cortex. These initial cortical and subcortical mappings create the foundations of the cognitive system and at the same time create the basic emotional structures. The elaboration of these emotional systems is the mechanism by which the human cortex relies not on genetics but on early experience to assess the world and develop appropriate ways to respond to it. As Jaak Panksepp explains:

> In the beginning, our higher neocortical brain, for all intents and purposes, is a *tabula rasa* of seemingly endless fields of self-similar columnar "chips" that are programmed by subcortical processes. Thus, practically everything that emerges in our higher neocortical apparatus arises from life experiences rather than genetic specializations.[6]

The challenge in understanding emotions that this chapter explores is that the flow of interoceptive information continues throughout life to signal important events, but at the same time, the *cortical* representations of emotional experience—with or without midbrain input—also come to play a central role in the shaping of emotional response.

The second idea grounding my approach extends the logic implicit in the central developmental role played by emotional substructures. These affective networks, in shaping the foundations of our cognitive systems, have a profound,

---

"Reconciling cognitive and affective neuroscience perspectives on the brain basis of emotional experience," *Neuroscience and Biobehavioral Review* 76 (2017):187-215.

[6] Jaak Panksepp, "The basic emotional circuits of mammalian brains: Do animals have affective lives?" *Neuroscience and Biobehavioral Review* 35 (2011), p. 1794.

pervasive impact: they "organize our internal world and our interactions with others."[7] That is, the most sophisticated forms of cortical Bayesian prediction are pointless if one outcome is not to be preferred over another. Where do these preferences come from? We tend to assume all sorts of "instinctive" preferences: for survival, for sex, for freedom from pain. These, however, do not just magically appear in the brain. The cortex is just a vast network of neurons; why should it care about the body? Without some form of intrinsic structural motivation, bodily pain, for example, is just more data. What teaches the cortex to care? What shapes the nature of that care? Neuroscientists increasingly have realized that the brain's affective systems provide the cortical structures for perception, cognition, and planning with the required intrinsic "motivation to move" that drives action.[8] Emotions thus are structurally the "intentional core of the organism."[9]

In this chapter I set out this contemporary view of the central structural role of emotion in how we see the world and live our lives. While I start from the early arguments of Jaak Panksepp that introduce the core affective systems, I then draw on the broad body of work that has accumulated since his early

---

[7] Daniel J. Siegel, "The Integrative Meaning of Emotion:"

> In essence, these ancient [core affective] systems organize our internal world and our interactions with others. Influencing both our subjective sense of being alive and the regulation of gene expression and neuroplasticity, these primes are fundamentally how we create ourselves—internally and interactively—across time. In this way, all animals have an emotional life as we link differentiated areas of our bodily and social world. (p. 96)

The essay is in Darcia Narvaez, Jaak Panksepp, Allan N. Schore, and Tracy R. Gleason, eds., *Evolution, Early Experience and Human Development: From Research to Practice and Policy* (Oxford: Oxford University Press, 2013)

[8] Stephen T. Asma and Rami Gabriel discuss how emotion provides an "organism's motivation to move itself" that serves as a biologically grounded form of Spinoza's conatus in *The Emotional Mind: The Affective Roots of Culture and Cognition* (Cambridge: Harvard University Press, 2019), p. 62.

[9] Rebecca M. Todd et al. argue that "Like other mammals, we are enmeshed in the world such that emotional outcomes are the central organizing feature of perception, attention, learning and memory, guiding awareness, and informing action." (See "Emotional Objectivity: Neural Representations of Emotions and Their Interaction with Cognition," *Annual Review of Psychology* 71 [2020], p. 43.) The phrase "intentional core" comes from Asma and Gabriel, *The Emotional Mind*, p. 27.

contributions in defining the field of affective neuroscience. I begin, as did Panksepp, with a set of midbrain nuclei that receive information from the body (interoceptive information), and then expand to explore how subcortical and cortical networks develop to map and elaborate the information as relayed by the midbrain and produce what we know as emotional experience.

A Brief Digression on Not Including a Survey of the History of Emotion

Many research articles on affective neuroscience begin by lamenting that what one means by "emotion" remains a matter of intense dispute.[10] Many then cite William James' famous argument that when we see a bear and run with a beating heart-rate, we then are afraid, based on the physiological response: "bodily changes follow directly the PERCEPTION of the exciting fact… and our feeling of the same changes as they occur IS the emotion."[11] Having cited James, the authors then go on to introduce the particular approach used in their own research. I initially planned to follow this pattern as well and provide the reader with a survey of the history of Western accounts of emotion. However, other books do this quite well.[12] And since my own thinking about the nature of emotion is more influenced by the classical Chinese tradition in any case, the

---

[10] Ralph Adolphs laments, "Affective neuroscience can be confusing when it fails to make distinctions between different aspects of affective processing. The titles of papers and discussions that authors give are often no help here either, since they frequently conflate different meanings of the term 'emotion'. The most common ambiguity is between 'emotion' as conceived above (the functional state) and its conscious experience, conceptualization or attribution." ("How should neuroscience study emotions? by distinguishing emotion states, concepts, and experiences" *Social Cognitive and Affective Neuroscience* 2017, p. 27)

[11] William James, "What is an Emotion?" *Mind* 9.34 (April 1884), pp. 189-90.

[12] In particular I recommend two related books. Jan Plamper's *The History of Emotions: An Introduction* translated by Keith Tribe (Oxford: Oxford University Press, 2015) describes not only the early history of emotions but also offers a not inappropriately skeptical overview of the formulations offered by psychologists, neurologists, and neuroscientists in the 20th and 21st centuries. Rob Boddice's *The History of Emotions* (Manchester: Manchester University Press, 2018) continues the task of outlining the past approaches and evolving methodologies of the study of the history of emotion and their relation to contemporary neuroscience.

Western history is not immediately relevant to the account in this chapter.[13] As I have noted, my story begins not with the earlier 20th century psychological accounts of emotion but with Jaak Panksepp's approach to affective neuroscience. In the presentation that follows, I introduce other major recent streams of thought from Paul Ekland's version of "basic emotions" to Lisa Feldman Barrett's contemporary constructive model to Antonio Damasio's description of an affectively charged "core self" and to a host of models in affective and cognitive neuroscience in between these positions. Even though I am a humanist, it is not clear to me that reviewing the historical antecedents to the positions—available elsewhere—will help clarify the particular models I explore.

The Three Tiers of Affective Organization

In this chapter I present three levels of neuronal systems that underlie emotional experiences.[14] The first layer is a network of brainstem and midbrain nuclei that respond to internal states and produce an array of neuromodulators (that enhance the effectiveness of neurotransmitters) and generate primary signals for *valence* (how good? how bad) and *salience* (how important). As we have seen for the attentional networks shaping the visual system, this layer works not simply through the localized activation of particular nuclei but through distinctive combinations that form functional networks. Nomenclature has long generated confusion in affective neuroscience: it is important not to think of these functional networks as producing *emotions* but as *dispositional matrices* that generate neuromodulatory signals to train the subcortical and cortical networks which then map the interaction of body states and external sensory inputs to produce what we usually think of as emotion.

These subcortical-cortical networks comprise the second level of the affective hierarchy and serve as a first-order representational system. In some cases, these second-layer networks form discrete units, as in the posterior insular cortex

---

[13] For an excellent discussion of the early Chinese traditions, see Curie Virág, *The Emotions in Early Chinese Philosophy* (Oxford: Oxford University Press, 2017).

[14] The basic insight is that of Jaak Panksepp as laid out in his *Affective Neuroscience* in 1998, but it has been adopted by a broad group of contemporary researchers.

that maps interoceptive data in a manner analogous to the visual system building its retinotopic maps of the visual data. But the neuromodulators of the brain stem and midbrain dispositional matrices also have a more pervasive effect on the postnatal structuring of sensory data in general. As we already have seen for the visual system, eye movement derives from feedback networks that calculate valence and salience. More broadly and especially before the higher-order systems of the prefrontal cortex develop, sensory perception and the sensory modeling of the external world—and its interactions with the interoceptive system—are strongly affected by modulation from brain stem / midbrain dispositional networks.[15] Thus, the neuronal networks that structure the sensory representations of the world are themselves not purely perceptual but also participate in the affective system.

The third level of the affective system builds upon the second level much as higher-order cognitive structures in semantic memory abstract regularities provided by the already ordered low-level pattern extraction capabilities of the perceptual systems. Since the initial sensory structures have an affective component, the structuring of higher-order semantic networks similarly incorporates structured affective information and participates in affective processing at a higher level of complexity and flexibility.[16]

---

[15] Eric E. Nelson, for example, concludes:

> This pattern of data suggests that what is most salient for the infant at a behavioral level corresponds to the functional systems that are undergoing construction in the nervous system and the information that is obtained behaviorally is integrated into the structure of the brain.

Eric E. Nelson, "The Neurobiological Basis of Empathy and Its Development in the Context of Our Evolutionary Heritage," in Darcia Narvaez, et al.., eds., *Evolution, Early Experience and Human Development*, p. 185.

[16] Rebecca M. Todd et al. in "Emotional Objectivity" argue:

> Emotional information not only engages the whole brain during an emotional experience but also tunes what we see of the world and helps determine how and what we learn and remember. Thus, rather than opposing emotion to cognition, we provide evidence that the primary function of brain-state representations is to produce unified emotion, perception, and thought (e.g., "That is a good thing") rather than discrete and isolated psychological events (e.g., "That is a thing. I feel good"). This understanding offers insight into how emotion operates as a

Jaak Panksepp provides a convenient summary of this model:

> Thus, it is useful to divide evolved brain functions in terms of primary-processes (tools for living provided by evolution), secondary-processes (the vast unconscious learning and memory mechanisms of the brain), and tertiary-processes (the higher order functions of mind permitted largely by the cortical expansions that allow many thought-related symbolic functions).[17]

However, one of the persistent complexities in the neuroscientific literature on emotion is the on-going interactions among these three levels. Panksepp represents this interaction as a set of *nested* systems:

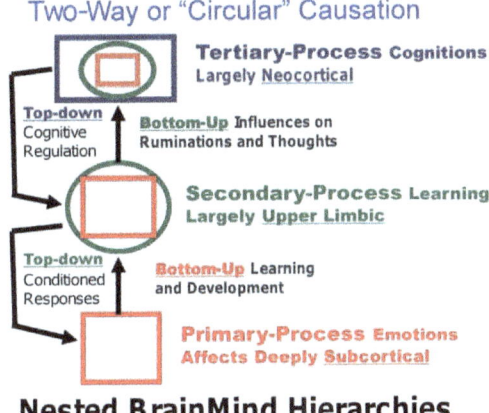

Figure 4.1: Affective Hierarchies[18]

After the tertiary systems mature, the primary brainstem and midbrain nuclei do not grow silent but continue to generate neuromodulators based on activations from the internal milieu, even if cortical and subcortical feedback connections

---

fundamental feature of cognition, ensuring that emotional outcomes are the driving principle of perception, thought, and action. (p. 27)

[17] Jaak Panksepp, "The basic emotional circuits of mammalian brains: Do animals have affective lives?" *Neuroscience and Biobehavioral Review* 35 (2011), p. 1792.

[18] This diagram appears in many accounts of Panksepp's model; this version comes from "Cross-Species Affective Neuroscience Decoding of the Primal Affective Experiences of Humans and Related Animals," *PLoS One* 6.9 (September 2011), p. 6.

now can inhibit those responses. Thus, emotional experience has three drivers in this model: brainstem interoceptive data, low-level subcortical-cortical mappings of an individual's history of correlation between interoceptive and sensory data, and, finally, cortical networks that build more complex memories, assessments, and plans integrating affect and experience. A related debate about the ongoing interactions between the levels of the affective system is the extent to which the sort of tertiary emotions that become conscious still depend not only on the second-level representations of the mapping of patterns of body-world interactions but also on the activation of the primary dispositional networks. To a significant extent, this question mirrors the questions of whether visual imagery requires feedback activation of networks in V1 and whether, as will be discussed later, access to the semantics of a verb requires the activation of the motor planning networks required to enact the motions that are part of the meaning of that verb. The depth of feedback activation is important in debates about emotion because "emotion" looks very different if one focuses on the tertiary structures for mapping the semantics of emotional experience or if one focuses largely on networks that include amygdala activation or if one stresses the yet deeper layers of hypothalamic activation.[19] The signatures of these different varieties of "emotion" also quite literally look very different when the fMRI data is introduced into the discussion.

To help disambiguate the levels of the affective system, I adapt Antonio Damasio's useful terminology of "dispositions," which structurally serve the same function as Panksepp's "primary-process emotions" that bootstrap and provide a basis for the later articulation of the emotional networks:

> We have inherited, from many prior species, abundant networks of dispositions that run our basic mechanisms of life management. They

---

[19] For example, David Sander, Didier Grandjean and Klaus R. Scherer in "An Appraisal-Driven Componential Approach to the Emotional Brain," *Emotion Review* 10.3 (July 2018), p. 234, argue:

> Importantly, current models of emotion do not argue that the bodily reaction always needs to occur within the body itself in order to be a component of emotion: the cerebral bodily map representation would be sufficient (e.g., an "as-if body loop" in Damasio's model), as has been suggested by embodiment theories of emotion.

include the nuclei that control our endocrine system and the nuclei that serve as the mechanisms of reward and punishment and the triggering and execution of the emotions.[20]

It is to these basic dispositional matrices that I turn next.

The Dispositional Matrices: Creaturely Foundations beyond Allostasis

Many accounts of emotion describe its basic role as a mechanism to maintain homeostasis or, in more recent accounts, to achieve allostasis. Lisa Feldman Barrett is a leading proponent of emotion as "interoceptive inference" to support allostasis and explains:

> brains did not evolve for rationality, happiness, or accurate perception. Brains evolved, in part, to efficiently ensure resources for physiological systems within an animal's body (i.e., its internal milieu) so that an animal can grow, survive, and reproduce. This balancing act is called allostasis….
>
> We hypothesize that emotions (as opposed to cognitions) are constructed when interoceptive sensations are intense or when the change in affect is large and foregrounded in awareness. This may help explain why, in mammals, the brain regions that are responsible for establishing and maintaining allostasis … are usually assumed to contain the circuits for emotion – they are important for regulating metabolism and energy expenditures associated with intense affect….[21]

Homeostasis is the process that maintains the stability of the physiological systems, and allostasis is the mechanism by which the animal responds to changes

---

[20] Antonio Damasio, *Self Comes to Mind: Constructing the Conscious Brain* (New York: Vintage, 2012), p. 144.

[21] Lisa Feldman Barrett and Ajay B. Satpute, "Historical pitfalls and new directions in the neuroscience of emotion," *Neuroscience Letters* (2017), pp. 12-13.

in the environment to assure homeostatic stability.[22] Barrett's argument is that emotions are part of the system that predicts how the world will affect the internal milieu and that therefore proactively plans responses to the anticipated change. In these models, maintaining homeostasis—or at least restoring it in response to the world—is the central goal. Thus, the focus is on the survival of the individual animal.

However, evolutionary natural selection produces drifts toward traits that assist the survival not of individuals but of genetically related populations. Social animals like humans have acquired phenotypical traits that integrate individuals into larger social structures. Humans, like other mammals, have social emotions that play crucial roles in the structuring of social order, emotions that range from impulses to nurture the young to the mix of cooperative play and aggressive competition that helps shape social hierarchies in larger groups.[23]

The distinctive feature of Jaak Panksepp's model for affective neuroscience is the centrality of the neurobiological systems that shape the developing brain in such a way that the social emotions arise as emergent properties of the maturing cortical structures. While the maintenance of homeostasis and the more active role of interoceptive active inference point to part of what emotion does and is, there is something more to emotion, given the behavioral requirements of more complexly evolved social animals like humans. Kenneth L. Davis and Christian Montag describe this larger domain that evolved for emotion:

> Primal emotions [my "dispositions"] and their accompanying affects appear to have acquired the capacity to move animals to action in ways that promoted their survival. Emotions prodded animals to explore for resources (SEEKING), compete for and defend those

---

[22] Anil K. Seth and Manos Tsakiris in "Being a Beast Machine: The Somatic Basis of Selfhood," (*Trends in Cognitive Sciences*, 22.11 [November 2018], pp. 969-81) give a good account of this model. Elizabeth S. Paul, et al., "Towards a comparative science of emotion: Affect and consciousness in humans and animals" (*Neuroscience and Biobehavioral Reviews* 108 [2020]:749-770) also provides a good comparative discussion of issues from an "appraisal" perspective.

[23] See, for example, the many essays in Darcia Narvaez, et al., eds., *Evolution, Early Experience and Human Development*.

resources (RAGE/Anger), escape from and avoid bodily danger (FEAR), and identify potential mates and reproduce (LUST). Then, mammals with their more social orientation acquired the motivational system for nurturing their offspring (CARE); the powerful separation distress system for maintaining social contact and social bonding (PANIC/Sadness); and the complex system stimulating especially young animals to regularly engage in physical activities like wrestling, running, and chasing each other (PLAY/Social Joy), which helps them bond socially and learn social limits and which seems to carry over into the "ribbing" and joking that continues to add fun in adulthood. Evolution has endowed mammalian brains with at least these seven primary-process emotional action systems, which serve as survival guides.[24]

The seven categories listed with capital letters are the "primordial anoetic affects" based in the brainstem and midbrain that ground what we know as emotion.[25] They are *anoetic* because the nuclei that generate these affects are not experience-dependent and do not draw on experience in their activations: they receive activations from the internal milieu and they fire. (Granted, later-developing cortical systems acquire the experience-dependent capability to either enhance or inhibit their firing, but the synaptic learning is in the cortical feedback connections to these nuclei.) Panksepp capitalized these categories to emphasize that these seven discrete systems are the *precursors* from which the corollary subcortical-cortical affective structures emerge. Influenced by these nuclei's neuromodulators, the cortical systems developmentally map the regularities in the interactions between the body and the world made available through the interoceptive, proprioceptive, and exteroceptive systems.[26]

---

[24] Kenneth L. Davis and Christian Montag, "Selected Principles of Pankseppian Affective Neuroscience," *Frontiers in Neuroscience* 12 (January 2019), p. 2.

[25] See Panksepp's tabular description of the levels of his model in "How Primary-Process Emotional Systems Guide Child Development: Ancestral Regulators of Human Happiness, Thriving, and Suffering" in Darcia Narvaez, et al., eds., *Evolution, Early Experience and Human Development*, p. 77.

[26] I take this terminology from Antonio Damasio, *Self Comes to Mind*, p. 80:

The Neurobiological Dispositional Networks

Panksepp describes subcortical nuclei involved in his seven dispositional matrices:[27]

Disposition Matrix / Manifestation	Key Brain Area	Key Neuromodulators
General Positive Motivation SEEKING / Expectancy	Nucleus Accumbens— VTA Mesolimbic and mesocortical outputs, Lateral hypothalamus, PAG	DA(+), glutamate, opioids (+), neurotensin (+), orexin (+), many other neuropeptides
RAGE / Anger	Medial amygdala to Bed Nucleus of the Stria Terminalis (BNST), Medial and perifornical hypothalamus to PAG	Substance P (+), Acetylcholine (Ach, +), glutamate (+)
FEAR / Anxiety	Central and lateral amygdala to medial hypothalamus and dorsal PAG	Glutamate (+), DBI, CRF, CCK, alpha-MSH, NPY
LUST / Sexuality	Cortico-medial amygdala, BNST, Preoptic hypothalamus, Ventromedial Hypothalamus, PAG	Steroids (+), vasopressin, oxytocin, LH-RH, CCK
CARE / Nurturance	Anterior Cingulate Cortex, Preoptic Area, VTA, PAG	Oxytocin (+), prolactin (+), dopamine (+), opioids (+/-)

I. Maps of the organism' internal structure and state (interoceptive maps)
   The functional condition of the body tissues such as the degree of contraction / distension of smooth musculature; parameter of internal milieu state.
II. Maps of other aspects of the organism (proprioceptive maps)
   Images of specific body components such as joints, striated musculature, some viscera
III. Maps of the world external to the organism (exteroceptive maps)
   Any object or event that engages a sensory probe such as the retina, the cochlea, or the mechanoreceptors of the skin.

[27] Jaak Panksepp, "Cross-Species Affective Neuroscience Decoding of the Primal Affective Experiences of Humans and Related Animals," *PLoS One* 6.9: e21236. doi:10.1371/journal.pone.0021236, p. 9, (doi:10.1371/journal.pone.0021236.g005) slightly modified. For PAG (periaqueductal gray), see p. 62, note 4.

PANIC /Separation	Anterior Cingulate Cortex, Preoptic Area, Dorsomedial Thalamus, PAG	Opioids (-), oxytocin (+), prolactin (-), CRF (+), glutamate (+)
PLAY / Joy	Dorsomedial diencephalon, Parafascicular Area, PAG	Opioids (+/-), glutamate (+), Ach (+), cannabinoids, TRH?

The lists of nuclei associated with each of the seven dispositional networks contain significant overlap. In particular, the periaqueductal gray (PAG) notably plays a role in all of these networks. However, since the time Panksepp first developed this list of dispositions and their anatomical key brain areas, network neuroscience has developed paradigms to help think about these overlaps. As discussed for the systems for control of attention, the focus has shifted away from functionality as strictly localized within single, distinct brain areas to the idea of functionally connected networks in which nuclei are nodes that depend on current activation patterns and participate in many different networks. Recent research has sought to trace these network connections for brainstem nuclei. The periaqueductal grey (PAG), that participates in all the dispositional networks, has attracted particular attention, although not for its role in what Panksepp calls "emotional affect" but in homeostatic affect, and in the processing of pain in particular.[28] Since the midbrain and brainstem nuclei produce their own distinctive neuromodulators, and a network of such nuclei defines the features of any given dispositional matrix, it is no surprise that these dispositional matrices rely to a significant extent on a shared set of neuromodulators. Thus, the PLAY matrix in Panksepp's table is not a single "thing" in its internal structure but a network of four main nuclei with specific connectivity that release a set of five neuromodulators that, when interacting with the developing cortex, produces

---

[28] Haley N. Harris and Yuan B. Peng, "Evidence and explanation for the involvement of the nucleus accumbens in pain processing," *Neural Regeneration Research* 15.4 (October 2019), p. 599. The legend for the figure reads "This figure represents both ascending and descending pain pathways, highlighting the nucleus accumbens and periaqueductal gray involvement, including excitatory (red) and inhibitory (blue) connections. ACC: Anterior cingulate cortex; Amy: amygdala; DH: dorsal horn of the spinal cord; Hab: habenula; Hyp: hypothalamus; Ins: insular cortex; LC: locus coeruleus; NAc: nucleus accumbens; NRM: nucleus raphe magnus; PAG: periaqueductal gray; RVM: rostral ventral medulla; S1: primary somatosensory cortex; Thal: thalamus."

the set of affective responses (at the subcortical-cortical level) and the behavior features that we call rough-and-tumble play. Panksepp and those who have followed his approach have stressed that among these seven dispositional matrices, SEEKING (and the dopamine system that embodies it) stands behind the other six as the primal mover, a "goad without a goal."[29]

Recent research has underscored the interconnected character of the systems that underlie basic dispositions. For example, the prairie vole has been the target of much study because of its strong pair-bonding. Neuroscientists have been attempting to understand the brain processes that lead to this behavior, which Panksepp identifies as one manifestation of the CARE matrix. As Panksepp proposed, they have discovered that in the vole, the dopamine system (Panksepp's SEEKING matrix) interacts with the oxytocin-producing nuclei with a particular developmental timing to produce the pair-bonding behavior.[30]

---

[29] As Asma and Gabriel describe it,

> SEEKING is often classed with the emotions, but it really is a master emotion or drive, a motivational system that organisms enlist in order to find and exploit resources in their environment. It energizes mammals to pursue pleasures or satisfactions, but it is not the same as pleasure. It is that growing, intense sensation of heightened attention and increasing feeling of anticipation—as if you are just about to scratch a powerful itch. Panksepp calls it a "goad without a goal," but the goad eventually does fasten onto specific goals. Its intrinsic aspect is promiscuous and flexible—motivating different pursuits at different times.

Stephen T. Asma and Rami Gabriel, *The Emotional Mind*, p. 61.

[30] Sue Carter and Stephen W. Porges describe the neuromodulator interactions that support pair-bonding in voles:

> The physiology of social bonding is also regulated by systems responsible for rewarding experience. For example, in socially naïve prairie voles, new pair bonds may form more readily when the partner is unfamiliar (i.e., not a member of the family), which reduces the potential for inbreeding and incest. In this context, pair bonds are reinforced by neural mechanisms that are shared with reward and pleasure. The dopamine system and, specifically, dopamine receptors in the nucleus accumbens have been implicated in pair bonding in prairie voles.... Dopamine-2 receptors appear to be important to the formation of pair bonds, while dopamine-1 receptors (activated by mating) may help to prevent the formation of new pair bonds while allowing the maintenance of previously formed social preferences. Oxytocin receptors are abundant in the nucleus accumbens and interaction between OT and dopamine in this brain region may help cement social bonds.

Parental behavior offers another example of the sort of interactions of brainstem and midbrain nuclei that generate dispositions important for the species.[31] "Maternal behavior" is not just one thing but a complex set of responses primed by a correspondingly complex network of subcortical nuclei.[32]

Sue Carter and Stephen W. Porges, "Neurobiology and the Evolution of Mammalian Social Behavior," in Narvaez, et al., eds., *Evolution, Early Experience and Human Development*, pp. 137-38.

[31] The coordination of the network of subcortical nuclei and their production of neuromodulators is complex, and an overview, based on research with mice, is worth citing in full to give a sense of how such a system works:

> Neurons in the MPOA [medial preoptic area] also project to the medial nucleus of the amygdala (MeA), where they act to inhibit competing social interactions and sustain maternal parenting behaviour. Thus, activity in the MPOA and its projections sensitizes the limbic network underpinning mammalian maternal care. The MPOA contains cells expressing various neurotransmitters and neuropeptides, and the diverse projections of these cells connect to multiple neural targets in the mammalian parenting network to support maternal behaviour. For instance, MPOA oestrogen-receptor-α-expressing cells send inhibitory projections to the VTA to inhibit non-dopaminergic cells and stimulate pup approach. Similarly, MPOA excitatory galanin-containing neurons project to multiple targets in the maternal brain, including the VTA, the periaqueductal grey (PAG) and the MeA, to orchestrate maternal care. Steroid-sensitive excitatory MPOA neurons that encode ethologically relevant social information project to VTA neurons, stimulating dopamine release to govern social approach behaviour. In addition to priming these MPOA neurons, oxytocin acts directly on neurons in the VTA to facilitate dopamine release in the nucleus accumbens, and oxytocin-primed synaptic plasticity in the amygdala supports the formation of social memories of the attachment target. Also included in the maternal caregiving circuit are parts of the mesolimbic dopamine network and several other regions rich in oxytocin receptors, including the bed nucleus of stria terminalis (BNST), the lateral septum and the ventral pallidum.

Ruth Feldman, Katharina Braun and Frances A. Champagne, "The neural mechanisms and consequences of paternal caregiving," *Nature Reviews: Neuroscience* 20 (April 2019), p. 209.

[32] For the male as well, there is a network of subcortical nuclei that primes paternal responses:

> Similar to mothers, specific pools of MPOA galanin-expressing neurons in the paternal brain project to inhibitory PAG neurons to promote pup grooming, to VTA neurons to increase approach behaviour and to the MeA to suppress competing social stimuli to help fathers focus on pups. Paternal behaviours are also influenced by communication with a female mate; lesions to the MPOA disrupt mate-dependent paternal behaviour in mice, and the expression of FOS is increased in the MPOA in male mice engaged in paternal care following ultrasonic vocalizations from the mother suggesting that male–female communication may

The interactions of midbrain affective networks produce the dispositions to bond in pairs and to act in maternal or paternal ways, but dispositions are not destiny. In humans, these subcortical systems entrain cortical networks for perception, cognition, memory, awareness, and emotion, but the ways that the cortical networks develop are experience-dependent and thus are profoundly influenced by the natures of the bodies housing the brains as well as the vicissitudes of experience. The Feldman group's diagram for the "paternal brain" (Figure 4.2) points to the range of distinct layers involved, including cortical regions for "embodied simulation" and "mentalizing" as well as the dispositional network of subcortical nuclei:

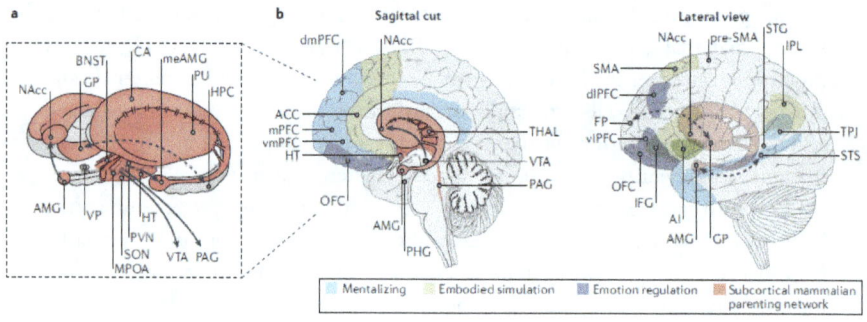

Figure 4.2: The Human Parenting Network[33]

Cortical Mapping of Affective Experience

"Embodied simulation" introduces the next layer in Panksepp's hierarchy: "secondary-process emotion" ("the vast unconscious learning and memory mechanisms of the brain"). Ruth Feldman, primarily interested in the neurobiology of attachment, defines "embodied simulation" as a more specific form of emotional mapping:

---

mediate some aspects of paternal caregiving and that the MPOA is involved in regulating this phenomenon.

Ruth Feldman, et al., "The neural mechanisms and consequences of paternal caregiving," p. 209.

[33] Ruth Feldman, et al., "The neural mechanisms and consequences of paternal caregiving," p. 210.

The second system underpinning human attachment is the 'embodied simulation/empathy network', including the insula, ACC, IFG [inferior frontal gyrus], IPL [inferior parietal lobe], and SMA [supplemental motor area]. Embodied simulation is an evolutionary-ancient mechanism, which, via automatic interoception and internal representations, recreates other's state in one's brain.[34]

For both Panksepp and Feldman, however, the second layer in emotional experience is a set of networks that both create cortical representations (i.e., memory maps) of the interaction between the primary affect-generating subcortical nuclei and the body and environment and develop feedback connectivity to regulate the primary layer. As Panksepp describes his secondary process emotional systems:

> At the secondary-process (noetic) level, our affective lives are parsed to fit and adapt to environmental opportunities and exigencies through the deeply unconscious neural mechanisms of learning and memory, which fill our memory banks with the grist for autonoetic consciousness—the tertiary process forms of both wise and unwise decision-making, evoked commonly in the service of our feelings.[35]

Many researchers propose models for emotion with a layer of cortical representations that map affective experience generated by subcortical nuclei. In these models, researchers mostly point to the same cortical regions—the insula, anterior cingulate cortex, the posterior parietal cortex, and orbitofrontal cortex—as well as the amygdala and hippocampus.

In *Self comes to Mind*, Antonio Damasio offers an important variation on the general model for mapping from brainstem to subcortical to cortical articulations of emotional structures. He proposes three layers to the self: a proto-self, a core self, and an autobiographical self. The proto-self has two layers of

---

[34] Ruth Feldman, "The Neurobiology of Human Attachments," *Trends in Cognitive Sciences* 21.2 (February 2017), p. 93

[35] Jaak Panksepp, "How Primary-Process Emotional Systems Guide Child Development: Ancestral Regulators of Human Happiness, Thriving, and Suffering," in Narvaez, et al., eds., *Evolution, Early Experience and Human Development*, p. 77

components that correspond to Panksepp's primary and secondary affective networks.[36]

> Brainstem level
> > *Interoceptive Integration*
> > > nucleus tractus solitarius
> > > parabrachial nucleus
> > > periaqueductal gray
> > > area postrema
> > > hypothalamus
> > > superior colliculus (deep layers)
>
> Cerebral Cortex level
> > *Interoceptive Integration*
> > > insular cortex
> > > anterior cingulate cortex
> > *External Sensory Portals*
> > > frontal eye fields
> > > somatosensory cortices

As Damasio explains the proto-self,

> "[The protoself] is an integrated collection of separate neural patterns that map, moment by moment, the most stable aspects of the organism's physical structure. The protoself maps are distinctive in that they generate not merely body images but also felt body images. These primordial feelings of the body are spontaneously present in the normal awake brain.
>
> The contributors to the protoself include master interoceptive maps, master organism maps, and maps of the externally directed sensory portals. From an anatomical standpoint, these maps arise both from the brain stem and from the cortical regions. The basic state of the protoself is an average of its interoceptive component and its sensory portal component. The integration of all these divers and

---

[36] Damasio, *Self Comes to Mind*, p. 203.

spatially distributed maps takes place by cross-signaling within the same time window. [Emphasis in the original][37]

Damasio's usage of the term "mapping" spans two distinct meanings. The first meaning refers to the stable self-organizing maps that structure neuronal *connectivity* throughout the brain but are especially crucial for thinking about perceptual systems, learning, and semantic memory. Damasio switches between calling these networks "maps" and "images."[38] The second meaning, found in the passage above, refers to the constantly shifting *patterns of activation* among neural networks that have been shaped by the first form of long-term mapping. Even in the second version of shifting activations, it is clear that while brainstem and mid-brain activations remain important, the cortical systems that integrate the interoceptive activations with sensory data as well as with additional learned cortical representations of interoceptive patterns provide the networks that largely shape the "core self."

Another important version of the hierarchy of networks that shape affective experience comes from within the world of cognitive neuroscience in the work of Ryan Smith and Richard D. Lane, who offer yet a more detailed set of hierarchical structures that preserves Panksepp's basic divisions between brainstem and mid-brain / sub-cortical and cortical / and higher cortical systems. They draw an explicit analogy in the perceptual hierarchy between the visual system and the affective system. For vision, they offer a model that follows the same outlines as the structures discussed in Chapter 3. That is, the "raw" information from the retina goes to the LGN and V1 which both transmit their first-order synthesis of the retinal data on to the higher-level visual areas (V2-V4) and receive feedback from them. The extrastriate visual cortices then pass their higher-order abstractions onto the inferotemporal cortex and anterior temporal lobe. These activations at every level are shaped by the attentional system and high-order processing of the visual data in working memory that links it with the multimodal pattern memory supporting cognition and planning.

---

[37] Damasio, *Self Comes to Mind*, pp. 201-02.

[38] See Footnote 27.

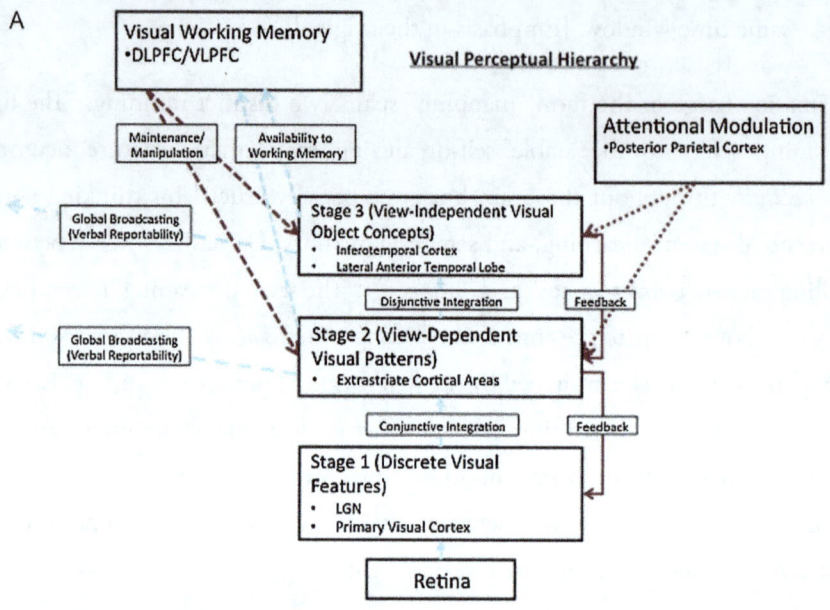

Figure 4.3a: The Visual Perceptual Hierarchy[39]

Smith and Lane argue that emotional perception operates in the same way. In this chart they collapse the source of activation into a simple "body state" that feeds into a mix of brain stem, mid-brain, and early cortical networks, particularly the posterior insula for interoceptive mappings and the somatosensory cortex that maps somatic data. As in the visual system, this early set of networks receive feedback from the higher-order cortical systems with which they are reciprocally connected.

This first stage of processing then goes to networks for higher order synthesis of patterns extracted from the interoceptive data with multimodal sense data as well as from the memory networks that preserve these patterns of interaction. The second-order processing is in the anterior insula, while the next

---

[39] Ryan Smith and Richard D. Lane, "The neural basis of one's own conscious and unconscious emotional states," *Neuroscience and Biobehavioral Review* 57 (2015), p. 8.

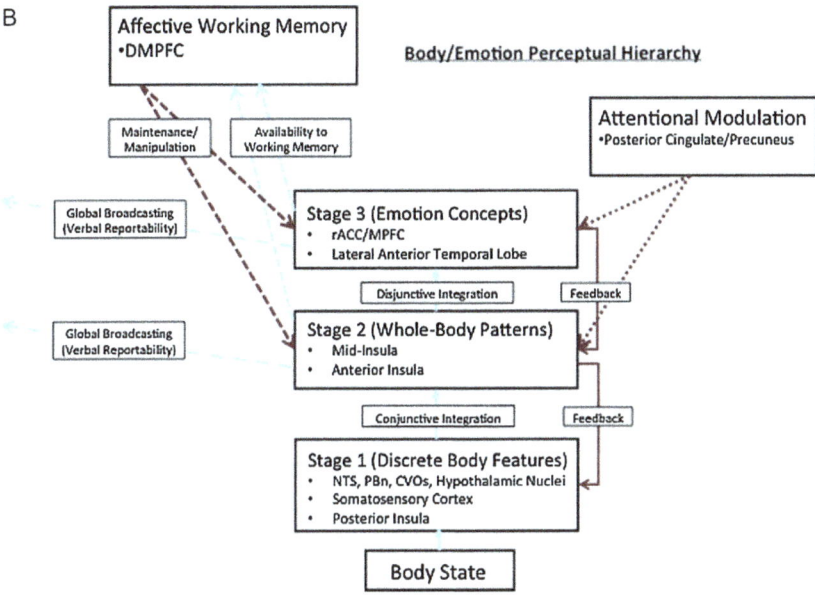

Figure 4.3b: The Body/Emotion Hierarchy[40]

layer involves the anterior cingulate cortex (ACC) and an increasingly broad set of networks for cognition, planning, and learning.

Smith and Lane's account deals with explicitly "emotional" responses rather than with the implicit and pervasive ways in which the affective systems shape memory, perception, and cognition more generally. But their models for the hierarchies involved in emotional perception, assessment, and response make it clear why simple fMRI scans fail to produce easily localized regions for the sorts of "basic emotions" proposed in earlier models. In particular, the ways in which the patterns of activation from Panksepp's set of dispositional networks are mapped into the posterior insula, thalamus, and the amygdala need not be simple. Instead, Smith and Lane describe a broadly distributed set of networks articulating ever higher-order mappings that abstract, represent, synthesize and process the responses of the brainstem/midbrain dispositional matrices.

---

[40] Smith and Lane, "The neural basis of one's own conscious and unconscious emotional states," p. 8.

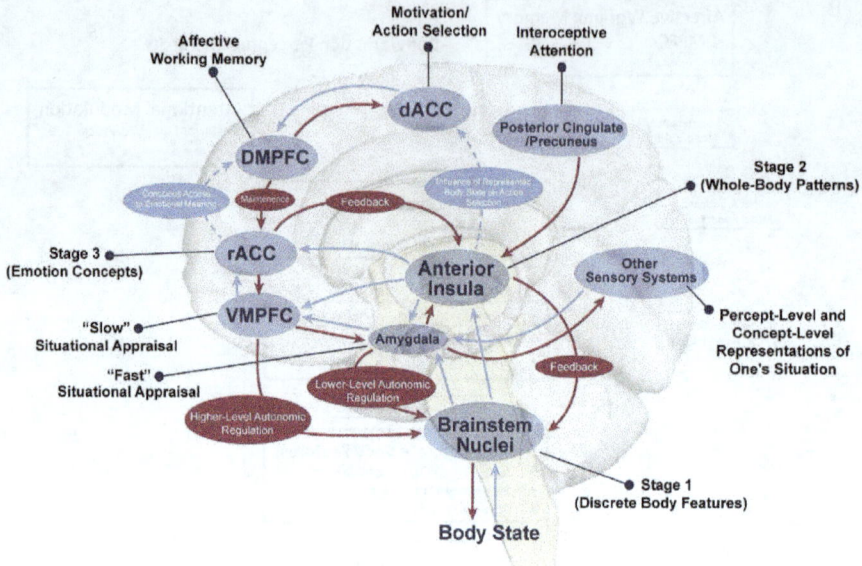

Figure 4.4: Affective Networks of the Brain[41]
dACC = dorsal Anterior Cingulate Cortex, DMPFC = Dorsomedial Prefrontal Cortex, rACC = rostral Anterior Cingulate Cortex, VMPFC = Ventromedial Prefrontal Cortex

Luiz Pessoa has been particularly important in stressing the complexly networked nature of emotional experience.

> I have proposed that the representation of emotion in the brain can be understood in terms of functionally integrated systems that involve large-scale cortical-subcortical networks that are sensitive to bodily signals. The high degree of signal distribution and integration in the brain provides a nexus for the intermixing of information related to perception, cognition, emotion, motivation, and action.[42]

The result of the way in which patterns of subcortical activations are integrated with other sources of sensory information and mapped into cortical regions interconnected with subcortical nuclei, is that, like visual information, they then

---

[41] Smith and Lane, "The neural basis of one's own conscious and unconscious emotional states", p. 17.

[42] Luiz Pessoa, "Understanding emotion with brain networks," *Current Opinion in Behavioral Science* 19 (2018 February): 19–25. Also see his "Emotion and the Interactive Brain: Insights from Comparative Neuroanatomy and Complex Systems," *Emotion Review* 10.3 (July 2018) 204–216.

become accessible for yet higher order integration into memory, the perceptual and cognitive networks, and planning. This is the level of abstraction and integration at which we experience emotion.

EXPERIENCED EMOTION: AFFECT IN HIGHER-ORDER CORTICAL PROCESSES

In the study of the higher-order cortical representation and processing of affect, there are two central questions. The first is how the higher order networks and their patterns of activation relate to the core dispositional networks that—through much mediation—ground emotional experience. The second is how environmental experience impinges on the higher order articulation of the core dispositional matrices.

Grounding Emotional Experience

Two quite different concerns shape inquiry into the relation of higher order emotional experience to the core dispositional networks that ground the affective system. The first is how to understand the nature of the emotional "concepts," through which we structure the vast range of highly differentiated forms of emotion that we experience in our own lives and read about in books. The second aspect of the connection of higher-order emotional experience to the core dispositional networks for which we must account is the "feltness" of feelings. In particular, much of the contemporary discussion of consciousness is about the sense of ownership of experience: that somehow, "I" see, that pain is my pain, and joy is my joy. The intuition, most strongly argued by Damasio but also shared by other scholars and scientists, is that this visceral sense of ownership is somehow connected to the systems that directly interact with activations from the viscera, that is, with the brainstem and midbrain nuclei and their secondary cortical elaborations. The question is *how* these visceral connections interact with higher-order cortical representations to shape the felt quality of conscious awareness.

Emotional Concepts

Emotional concepts are of particular concern to researchers who approach emotional experience through some form of constructionist model:

> According to constructionist views, core affect is transformed into an emotion experience during situated conceptualization when it is linked to the context and made meaningful as a specific emotion using conceptual knowledge about emotion categories.[43]

Lisa Feldman Barrett has developed an important variation of this constructionist approach by integrating the perspective of predictive coding into the appraisal process. The first step is the construction of the emotion:

> We hypothesize that emotions (as opposed to cognitions) are constructed when interoceptive sensations are intense or when the change in affect is large and foregrounded in awareness.[44]

However, with the idea of interoceptive inference, Barrett stresses that the brain does not simply passively respond to data but actively predicts the probable cause of the intense interoceptive sensation and develops plans to respond to that predicted cause. Cortical regions responsible for the regulation of the internal milieu provide feedback to the midbrain and brainstem nuclei.[45] The major

---

[43] Jennifer K. MacCormack and Kristen A. Lindquist, "Bodily Contributions to Emotion: Schachter's Legacy for a Psychological Constructionist View on Emotion," *Emotion Review* 9.1 (January 2017), p. 37.

[44] I also cite and provide a longer extract of Barrett's discussion on page 122. Lisa Feldman Barrett and Ajay B. Satpute, "Historical pitfalls and new directions in the neuroscience of emotion" *Neuroscience Letters* (2017), p. 6.

[45] Barrett explains:

> Limbic cortices, such as the anterior cingulate cortex and the ventral portion of the anterior insula (aINS), allostatically control physiology by relaying descending prediction signals to the internal milieu via a system of subcortical regions, including the central nucleus of the amygdala, the ventral and dorsal striatum, and the central pattern generators across hypothalamus, the parabrachial nucleus, periaqueductal grey, and the solitary nucleus.

Barrett, "The theory of constructed emotion: an active inference account of interoception and categorization," p. 9.

difference so far in this model is that Barrett specifically privileged allostatic goals over the more complex dispositional dynamics in Panksepp's approach. Interoceptive prediction itself is at the level of higher-order cortical structures in which the specific causal logic of brainstem activation (Panksepp or Barrett) is less relevant than the fact that there simply *is* as causal logic upon which inferential networks can be built. The task of these networks is precisely to model that logic. Barrett stresses that interoceptive inference requires that this cortical model be built and that it be constantly updated to improve the accuracy of prediction:

> The mechanistic details of predictive coding provide yet another deep insight: a brain implements its internal model with 'concepts' that 'categorize' sensations to give them meaning.[46]

Barrett here refers to "concepts" in quotes to warn her readers that such concepts are not the stuff of propositional logic and the older serial-computer model of the brain.

> In the language of the brain, a concept is a group of distributed 'patterns' of activity across some population of neurons. Incoming sensory evidence, as prediction error, helps to select from or modify this distribution of predictions, because certain simulations will better fit the sensory array (i.e., they will have stronger priors), with the end result that incoming sensory events are categorized as similar to some set of past experiences.[47]

---

[46] Lisa Feldman Barrett, "The theory of constructed emotion," p. 7.

[47] Lisa Feldman Barrett, "The theory of constructed emotion," p. 9. William A. Cunningham, Kristen Dunfield, and Paul E. Stillman offer the idea of attractor states (discussed in Chapter One) as the functional equivalent to Barrett's "concepts":

> In cognitive processing, the units of networks tend to gravitate toward stable patterns of activation called *attractor states*. An attractor state is a state of activation that is probable given a diverse array of neuronal activations. In other words, given different inputs, the network will settle on a single internal representation. (p. 172)

In their model, attractor state dynamics provide the mechanism by which Barrett's "interoceptive inference" proceeds. Their focus in particular is on the iterative processes by which inference settles into solutions and on the challenges in reaching solutions when the

In Barrett's view,

> A concept is a collection of embodied, whole brain representations that predict what is about to happen in the sensory environment, what the best action is to deal with impending events, and their consequences for allostasis (the latter is made available to consciousness as affect). Unpredicted information (i.e., prediction error) is encoded and consolidated whenever it is predicted to result in a physiological change in the state of the perceiver (i.e., whenever it impacts allostasis).[48]

As noted before, Barrett's model assumes that the brainstem and midbrain nuclei that receive information from the internal milieu transmit that data to subcortical and cortical networks as an accurate reflection of the state of the internal milieu and therefore of allostatic requirements. Panksepp's account of the dispositional networks complicates this idea of direct transmission (although allostatic information certainly is part of what is transmitted) with the introduction of biases in responses, shaped by evolution, that produce more complex forms of behavior less directed at the survival of the individual than at the success of the larger group. Still, from the perspective of that larger theory, this change is no more than a swapping out of computational modules whose patterns of response the cortical structure of emotional concepts must model. Once the cortical networks for constructing emotional concepts begin to develop, they

---

system encounters external data that either present entirely new patterns or do not conform to expectations:

> Entropy [a measure of the number of possible network states] can be difficult to reduce for a number of reasons, but two important sources of persistent entropy are either novel stimuli that do not settle easily into a preset attractor state, or violations of expectations that not only fundamentally alter one's current representation but also disrupt other representations in the network (or at least other layers of the network). (p. 175)

I cite their discussion because it remains important to keep in mind the systemic logic of the neural networks that structure Barrett's system of "concepts." See William A. Cunningham, Kristen Dunfield, and Paul E. Stillman in "Affect Dynamics: Iterative Reprocessing in the Production of Emotional Responses" in Lisa Feldman Barrett and James A. Russell, eds., *The Psychological Construction of Emotion* (New York: Guilford Press, 2015)

[48] Lisa Feldman Barrett, "The theory of constructed emotion," p. 12.

grow in depth and complexity based on experience, experience that includes not just physical encounters but also learning based on language and other symbolic systems. Concepts build upon themselves in the construction of the self-organizing maps that structure the emotional domain. As Barrett et al. note, "Even a single prediction can be constructed via conceptual combination from past experiences belonging to different emotion categories."[49] Still, the highly articulated emotional structure of adult experience remains rooted in connectivity between the second-order cortical representational system, the first-order cortical mappings of the interoceptive information, and the midbrain and brainstem networks.

The Feeling of Emotion

Feeling is at the core of consciousness and selfhood but is a mystery. Christof Koch, one of the major scholars in the search for neural correlates of consciousness (NCC), for example, directly equates feeling with experience: for Koch, consciousness is simply "any experience, from the most mundane to the most exalted…. As I use it, any feeling is an experience. Collectively taken, then, consciousness is lived reality."[50] Michael Graziano, who also works on the neuroscience of consciousness, states the problem thus:

> You can connect a digital camera to a computer and program the system to process the incoming visual information. The computer can extract color, shape, and size, and it can identify objects. The human brain does something similar. The difference is that people have a subjective experience of what they see. We don't just register the information that the object is red; we have an *experience* of redness. Seeing *feels* like something. A modern computer can process a visual

---

[49] Katie Hoemann, Maria Gendron, and Lisa Feldman Barrett "Mixed emotions in the predictive brain," *Current Opinion in Behavioral Science* 15 (June 2017), p. 53.

[50] Christof Koch, *The Feeling of Life Itself: Why Consciousness is Widespread but Can't Be Computed* (Cambridge: MIT Press, 2019), p. 1.

image, but engineers have not yet solved how to make the computer conscious of that information.[51]

Richard Lane and Ryan Smith, drawing on the perspectives of cognitive neuroscience (CN), in conversation with Jaak Panksepp and Mark Solm offer a more focused model, not of experience generally but of "phenomenological emotional experience," that is, emotions as we experience them in conscious awareness:

> Therefore, while subcortical emotion circuits are important for generating these broad multifaceted emotional reactions, from the CN perspective, a phenomenological emotional experience involves considerable cortical contributions associated with representation of bodily sensations and their conceptual meaning, as well as cortical mechanisms associated with affective working memory and selective broadcasting.[52]

Anil K. Seth, another major researcher exploring the neuroscience of consciousness, adds selfhood, a key component of conscious experience, to the discussion. Seth proposes, in a manner analogous to Lane and Smith, that selfhood derives partly from the *representations* of the homeostasis of the affective system and partly from the higher cortical elaborations of first-person experience:[53]

> Selfhood is a constellation concept that involves not only representtation and control of physiological homeostasis, but also the experience of owning and identifying with a particular body, the

---

[51] Michael S. A. Graziano, *Rethinking Consciousness: A Scientific Theory of Subjective Experience* (New York: W. W. Norton & Company, 2019), p. 5.

[52] Lane and Smith in Jaak Panksepp, Richard D. Lane, Mark Solm, Ryan Smith, "Reconciling cognitive and affective neuroscience perspectives on the brain basis of emotional experience," *Neuroscience and Biobehavioral Review* 76 (2017), p. 192.

[53] Seth notes:

> There is also increasing focus in consciousness research on experiences of *selfhood*, which encompass basic experiences of embodiment and body ownership, experiences of volition and agency, as well as 'higher' aspects of selfhood such as episodic memory and social perception.

Anil K. Seth, "Consciousness: The last 50 years (and the next)," *Brain and Neuroscience Advances* 2 (2018), p. 4.

emergence of a first-person perspective, intention and agency, and metacognitive aspects that relate to the subjective 'I' and the narrative linking of episodic memories over time[54]

No neuroscientist, however, has worked so intently on the centrality of emotions in the deeply interconnected problems of feelings, the self, and consciousness as Antonio Damasio. As noted earlier, Damasio proposes three layers of "self:" the proto-self, the core self, and the autobiographical self. As discussed above, the proto-self essentially corresponds to the first two levels of affective networks in Panksepp's model and includes the brainstem/midbrain nuclei and the cortical areas that construct stable maps of the activations from the brainstem/midbrain networks. The core self, in contrast, is more of a state of brain activation than a stable set of networks. Damasio speaks of "changes" in the proto-self that come "from its moment-to-moment engagement as caused by any object being perceived." The core self arises as a response to these transient changes in the proto-self:

> Changes in the protoself inaugurate the momentary creation of the core self and initiate a chain of events. The first event in the chain is a transformation in the primordial feeling that results in a "feeling of knowing the object," a feeling that differentiates the object from other objects of the moment. The second event in the chain is a consequence of the feeling of knowing. It is a generation of "saliency" for the engaging object, a process generally subsumed by the term *attention*, a drawing in of processing resources toward one particular object rather than another. The core self, then, is created by linking the modified protoself to the objects that caused the modification, an object that now has been hallmarked by feeling and enhanced by attention.[55]

---

[54] Anil K. Seth, "Interoceptive inference, emotion, and the embodied self," *Trends in Cognitive Sciences*, Vol. 17, No. 11 (November 2013), p. 565.

[55] Damasio, *Self Comes to Mind*, p. 215.

Finally, Damasio's "autobiographical self," as the name suggests merges the activations of the current state of the core self with the networks of autobiographical memories. "The autobiographical selves are autobiographies made conscious. They draw on the entire compass of our memorized history, recent as well as remote."[56] Damasio proposes an outline for how this autobiographical self arises:

> First, a substantial set of defining biographical memories must be grouped together so that each can be readily treated as an individual object. Each such object is allowed to modify the protoself and produce its pulse of core self, with the respective feelings of knowing and consequent object saliency in tow. Second, because the objects in our biographies are so numerous, the brain needs devices capable of coordinating the evocation of memories, delivering them to the protoself for the requisite interaction, and holding the results of the interaction in a coherent pattern connected to the causative objects.... From a neural standpoint the coordinating process is especially complicated by the fact that the images that constitute an autobiography are largely implemented in the image workspace of the cerebral cortex, based on recall from dispositional cortices, and yet, in order to be made conscious, these same images need to interact with the protoself machinery, which, as we have seen, is largely located at brain-stem level.[57]

Damasio is fundamentally concerned with the "feeling" of experience generally and argues that the first-person quality of all such experience—not just that of "emotional" experience—derives from interactions with the brainstem structures and the cortices that map them.[58] In Damasio's account, the brainstem structures tend to be recruited by the cortical networks for object recognition and salience-

---

[56] Damasio, *Self Comes to Mind*, p. 222.

[57] Damasio, *Self Comes to Mind*, p. 222.

[58] Damasio does specifically discuss emotions in the chapter "Emotions and Feelings" in *Self Comes to Mind*. Although the chapter explores the particular qualities of emotional experience, his arguments crucially set up the arguments he needs to move from the protoself through the core self to the autobiographical self in later chapters.

assessment discussed in Chapter 3. The images (representations) of patterns of brainstem activation linked to cognized "objects" through experience give the experience its felt quality, but the recruitment of the networks of autobiographical memory in the construction of Damasio's autobiographical self make the experience "mine" in a complementary way. I find Damasio's synthesis particularly compelling for his account of "dispositional memory," but it should be clear that the broad outlines of his three-tiered model are consistent with those of the other researchers I have presented. Future work will more thoroughly integrate predictive processing and network neuroscience paradigms into our understanding of emotional experience and provide greater detail about the specific functional connectivity underlying the emotional networks. But the basic models, like that of Damasio, are sufficiently mature that humanists should give them careful consideration.

The Experiential Articulation of Emotions and the Role of Humanistic Inquiry

The cortically articulated, experientially shaped third tier of emotional organization in the brain links the neuroscientific paradigms to the world of humanistic discourse. Contemporary neuroscientific models stress that emotional organization within an individual is, by its very design, profoundly shaped by the vicissitudes of experience as well as by the structures of the encountered world. Emotional experience as an emergent phenomenon has a particularity and a history not easily, or even not possibly, predicted by what we know of the neuroscientific substructures alone. The emergent properties are not those of an individual but of the interactions between the individual and material, social, and cultural milieus that are themselves shaped by the constraints of an evolved human affective architecture. To understand the patterns of emotional experience, then, neuroscience is not enough; it requires the breadth and depth of humanistic reflection to complement the explorations of the neuronal substructures. This conclusion is built into the neuroscience.

From Psychological Construction to the History of Emotions

Lisa Feldman Barrett and her colleagues are clear that, in the end, the individual emotional dynamics that they can study with their methods incorporates a world of external structures that must be understood if we are truly to understand emotional experience even at the neuronal level.[59] Culture plays a central role in their model:

> Linked to variation within the conceptual system for a given emotion category is variation in the recurring situations that people find important and meaningful for a given emotion within a cultural context. If the conceptual system for emotion is constituted out of past experience, and if past experience is largely structured by people within a cultural context, then both emotion categories that develop and the population of instances within each category will be culturally relative.[60]

---

[59] They argue:

> The brain state corresponding to an emotional episode is not just whatever happens in the body, in the subcortical neurons responsible for fighting, fleeing, freezing, or mating, and so forth, or in the brain regions that represent or regulate the body (e.g., the insula, amygdala, and orbitofrontal cortex). Our hypothesis is that the brain state for an emotional instance is a representation of the state of affairs in the world in relation to that physical state; both sensations from the world and from the body are made meaningful by information stored in the brain from past instances, and so include a neural representation of whatever portion of that information is being used. Thus, the second source of variation within an emotional category derives from the conceptual knowledge that it contains.

Lisa Feldman Barrett, Christine D. Wilson-Mendenhall, and Lawrence W. Barsalou, "The Conceptual Act Theory" in Barrett and Russell, eds., *The Psychological Construction of Emotion*, p. 93.

[60] Lisa Feldman Barrett, et al., "The Conceptual Act Theory" in Barrett and Russell, eds., *The Psychological Construction of Emotion*, p. 97.

Even researchers who focus largely on the neurophysiology of emotion stress the crucial structuring role of experiential patterns incorporated from the external world:

> [A] psychological constructionist approach emphasizes not only that the mind is situated in the body, but also that conceptualization of core affect is situated with a social, cultural, historical, and current perceptual context (i.e., "situated conceptualization"; ...) Therefore, the conceptual features of an emotion will have considerable similarities across individuals within a specific cultural context due to

The "cultural turn" introduced into contemporary affective neuroscience has been met within the humanities by scholars who have focused on historical patterns of emotional experience in the emerging field of the history of emotion. There is an opportunity here for a significant conversation across disciplines. While Jan Plamper has outlined a history of biologically determinist accounts of emotion in *The History of Emotion: An Introduction*, the role for society and culture envisioned by contemporary neuroscience aligns closely with Rob Boddice's sense of the possibilities for collaboration:

> The history of emotions, working in tandem with psychologists of a different stripe, as well as neuroscientists, anthropologists, linguists and philosophers, have the opportunity, if not the responsibility, to disrupt the landscape of crude essentialism and emotional reductionism.[61]

The concepts of "emotional communities" and "emotional regimes" developed in the rapidly emerging field of the history of emotion capture the role of communities of shared emotional understanding and practices that are a part of a historically situated social and cultural milieu. "Emotional communities" give specific embodiment to the sorts of external regularities that inform the articulation of higher-order emotional structures in the cortex.[62] For example, Barbara Rosenwein describes "emotional communities":

> shared cultural influences (…). On the other hand, individuals with different personal histories (e.g., those from stable vs. unstable households of origin) could have different habitually elicited patterns of core affect, or different tendencies to attend to and conceptualize particular patterns of core affective change. Again, contextual details are important in the experience of an emotion; they are not experimental noise.

Ian R. Kleckner and Karen S. Quigley, "An Approach to Mapping the Neurophysiological State of the Body to Affective Experience" in Barrett and Russell, eds., *The Psychological Construction of Emotion*, p. 290.

[61] Rob Boddice, *The History of Emotions*, p. 38.

[62] See Boddice's discussion of Barbara Rosenwein's "emotional communities" and William Reddy's "emotional regimes" in *The History of Emotion*, pp. 62-83. Rosenwein offers her own comparison of the two concepts in Barbara H. Rosenwein and Riccardo Cristiani, *What is the History of Emotions* (Cambridge: Polity Press, 2018), and both Reddy and Rosenwein have implemented their approaches, Reddy in *The Making of Romantic Love: Longing and Sexuality*

> Emotional communities are groups—usually but not always social groups—that have their own particular values, modes of feeling, and ways to express those feelings. Like "speech communities," they may be very close in practice to other emotional communities of their time, or they may be quite unique and marginal. They are not bounded entities." Indeed, the researcher may define them quite broadly—upper-class English society in the nineteenth century, for example—or quite narrowly.... More narrowly delineated communities allow the researcher to characterize in clearer fashion the emotional style of the group. Larger communities will contain variants and counter-styles—"emotional subcommunities" if you will.[63]

However, the world of external patterning that is internalized in the construction of the experience-dependent cortical structures for emotion includes more than just people. Boddice correspondingly expands the scope of inquiry in the history of emotions:

> If one of the most basic premises of the study of emotions in the past is that emotional experience is constructed in context, then not only the people, but the *things* of that context, and the spaces themselves, become important. What we feel is often inextricably bound up with the things we feel *about*, and those things—animate and inanimate—derive their meanings and importance from the cultural web in which they are produced and found.[64]

The confluence of the models developed in contemporary affective neuroscience and the paradigms of the history of emotion should have a profound effect on the way we think not only about emotional experience but about all the daily

---

in Europe, South Asia & Japan, 900-1200 ce (Chicago: University of Chicago Press, 2012), and Rosenwein in *Generations of Feeling: A History of Emotions, 600-1700* (Cambridge: Cambridge University Press, 2016). Still, Boddice's conclusion largely holds: "In sum, anything that looks like an emotional community in Rosenwein's terms is probably also an emotional regime in Reddy's. The differences attributed to them pale in comparison to the similarities." (p. 80).

[63] Rosenwein, *Generations of Feeling*, p. 3.

[64] Boddice, *The History of Emotion*, p. 38.

elements of our experience of ourselves and the world. Boddice underscores how these models should cause us to rethink the character of our simultaneously biological and historical being in the world:

> [I]f the brain is imagined as a biological organ that is in part *made* by the world it is in, then two things happen: the world takes on a remarkable new importance in our understanding of who we are, and the brain becomes historical because it is a contextual object.[65]

The neuroscience leads us to this conclusion: far from being reductive or deterministic, it opens out, gives weight, complexity, and history to the dynamics of the development of the self.

The cultural and historical contexts of the experience-dependent maturation of the affective structures that give meaning to the self and the world also greatly complicate the challenges of our understanding of both ourselves and one another. We cannot fully master our own past or that of any other.[66] Emotions also change: we cannot rely on our own naïve, immediate responses, since they are as historically (and biographically) bound as those of the person whom we seek to understand. Thought systems matter; moral and religious convictions matter; they have substantive weight in our emotional lives.

---

[65] Boddice, *The History of Emotion*, p. 143

[66] Boddice provides a vignette illustrating this challenge:

> I remember coming to the end of a long series of letters by George John Romanes, the Darwinian disciple. As the brain tumour that would ultimately kill him progressed, so his correspondence became tinged with a sense of his own mortality. As his handwriting faltered, he apologized to his correspondents, explaining the frustration of his failing eyesight and motor skills. One comes to know a subject. To read his decline in his own hand was moving, deeply affecting and not a little difficult. Nonetheless, my emotional response to these letters had to be packed away in the course of analysis, because I could not assume that my response was, for all its empathy, hitting the right notes. To understand what a contextual empathy might have been like, I had to add back in the context of religious doubt that nagged Romanes' life, as well as the polarizing convictions of his correspondents in the worlds of science and religion, respectively. I had to understand the context, meaning, and reception of death and its correlation with age, as well as Romanes' own appreciation of these things.

Boddice, pp. 126-27.

Boddice's assertion of the centrality of the history of emotions in exploring the historical logic of the construction of meaning in human experience quite remarkably parallels both the neuroscientific arguments and a very differently derived approach to general hermeneutics that I find compelling. Wilhelm Dilthey (1833-1911) sought a way to ground the possibility of historical understanding and in the end turned to three interrelated ideas—lived experience, life expressions, and objective spirit—as the objects of historical inquiry and of human understanding. Lived experience is experience informed by a structure of human purposes.[67] If one picks up a hammer or writes a sonnet, for example, the humanly intelligible meanings of these actions are in the purposes expressed in the actions. However, as expressions of lived experience, the actions are parts of an individual's larger structure of purposiveness that reaches deep within.

> An expression of lived experience can contain more of the nexus of psychic life than any introspection can catch sight of. It draws from depths not illuminated by consciousness. But at the same time, it is characteristic of an expression of lived experience that its relation to the spiritual or human content expressed in it can only be made available to understanding within limits. Such expressions are not to be judged as true or false but as truthful or untruthful.[68]

---

[67] Dilthey develops an account of purposiveness:

> To the extent that the parts [of the experiential nexus] are connected structurally so as to link the satisfaction of the drives and happiness and to reject suffering, we call this nexus purposive. It is solely in psychic structure that the character of purposiveness is originally given, and when we attribute this to an organism or to the world, this concept is only transferred from inner lived experience. Every relation of parts to a whole attains the character of purposiveness from the value that is realized in it. This value is experienced only in the life of feelings and drives.

Wilhelm Dilthey, *Understanding the Human World, Volume II of Selected Works*, edited by R.A. Makkreel and F. Rodi (Princeton: Princeton University Press, 2010), p. 178.

[68] Wilhelm Dilthey, *The Formation of the Historical World in the Human Sciences, Volume III of Selected Works*, edited by R.A. Makkreel and F. Rodi (Princeton: Princeton University Press, 2002), p. 227.

Finally, the structuring of purposiveness is historically shaped. Goals, desires, and fears, as well as the modes of responding to those goals, desires, and fears, take shape within the historical milieu of objective mind. Dilthey explains:

> I have shown how significant the objective mind is for the possibility of knowledge in the human studies. By this I mean the manifold forms in which what individuals hold in common have objectified themselves in the world of the senses. In this objective mind, the past is a permanently enduring present for us. Its realm extends from the style of life and the forms of social intercourse to the system of purposes which society has created for itself and to custom, law, state, religion, art, science and philosophy. For even the work of genius represents ideas, feelings and ideals commonly held in an age and environment. From this world of objective mind the self receives sustenance from earliest childhood. It is the medium in which the understanding of other persons and their life-expressions takes place: For everything in which the mind has objectified itself contains something held in common by the I and the Thou. Every square planted with trees, every room in which seats are arranged, is intelligible to us from our infancy because human planning, arranging and valuing—common to all of us—have assigned a place to every square and every object in the room. The child grows up within the order and customs of the family which it shares with other members and its mother's orders are accepted in this context. Before it learns to talk, it is already wholly immersed in that common medium. It learns to understand the gestures and facial expressions, movements and exclamations, words and sentences, only because it encounters them always in the same form and in the same relation to what they mean and express. Thus the individual orientates himself in the world of objective mind.
>
> This has an important consequence for the process of understanding. Individuals do not usually apprehend life-expressions in

isolation but against a background of knowledge about common features and a relation to some mental content.[69]

Dilthey, that is, proposed a framework for the understanding of the construction of human meaning that encompasses the terms explored in this chapter. Humans share basic dispositions that find their individual articulation through engagement with a world of people and things. The articulated structure is beyond the reach of self-reflection but is manifest in action, and we can hope to understand those actions when situated within specific social and cultural contexts. Neuroscience looks to the inward dispositional structures and their experiential articulations through developmental processes. The history of emotions focuses on the patterns of contextual elements informing emotional structuring, and Dilthey opens out the project to the wider humanities.

My point is not that we are merely rediscovering old conceptual frameworks but that contemporary paradigms in the neuroscience of emotion offer a significant (empirical rather than ontological) grounding for the understanding of human experience. Thus, the neuroscientific models for emotion matter profoundly to humanistic inquiry. Yet while an awareness of the subcortical matrices for dispositions, their affective maps, and the cortical structures that elaborate our emotional organization is crucial, we also need to explore the specific, distinctive patterns of human brain development that strongly shape the emergent cortical networks behind both affective life and the structuring of the self. These patterns are the topic of the next chapter.

---

[69] Wilhelm Dilthey, *Draft for a Critique of Historical Reason*, translated in Kurt Mueller-Vollmer, ed., *The Hermeneutics Reader: Texts of the German Tradition from the Enlightenment to the Present* (New York: Continuum Press, 1985), p. 155. Cf. Wilhelm Dilthey, *The Formation of the Historical World in the Human Sciences, pp. 229-230.*

Chapter Five

Building a Brain, Shaping a Self

The self and the world in which it participates emerge simultaneously in the human brain as it discovers patterns encountered in early experience. This process takes time, and the brain matures slowly and in stages. As the visual system has shown us, constructing the internal neuronal models for complex experiential patterns must proceed one structural level at a time: the visual cortex first extracts line segments and then goes on to create representational spaces for ever more complex objects. The building of the neural networks of the brain follows the same sequential logic, first capturing the basic components for sensory experience, then, based on these primary networks, building layer upon layer of neuronal systems to capture ever more complex patterns in the representations of the world and the body, to integrate ever more complex (neuronally represented) interactions between the world and the body, and to construct increasingly flexible plans for response. Interacting neuronal systems build upon themselves to create ever more powerful modes of grasping and creating meaning in the world. This logic holds from the organization of the networks for extracting the most basic low-level regularities in visual and acoustic input to identifying meaningful objects in the world, to learning language, to growing into one's place in one's family, society, and culture.

In this chapter I trace the sequential, hierarchical logic by which brains grow into the world, beginning with the basic genetically encoded processes that direct the development of the gross architecture and connectivity of the brain during gestation and early infancy. I then set out the complex coordination of progressive synaptic reordering and myelination in which subcortical neuromodulators crucially guide the emerging neuronal structures that take shape through our initial engagements with the world. I next examine this developmental sequencing from the perspective of how the subcortical mediations lead at first to attachment to a caregiver and then to ever-widening explorations as the infant's range of action grows. In concluding the chapter, I consider the nature of the self that takes shape within this developmental matrix.

In the Beginning: The Early Development of the Human Brain

In the human fetus, the brain develops through a sequence of genetically controlled cellular transformations. This remarkable process is worth exploring in some detail because it helps clarify both the centrality of the unspooling of genetically encoded information in early development as well as the limitations of that information.[1] In humans, as in all altricial animals, by the time the child is born, the genetically organized large-scale neuronal structures are ready to take the next step in their development: these initially epigenetically orchestrated structures begin to construct more specialized neural networks through the incorporation of new information from the body and the environment.

The very earliest stage begins simply with the cellular differentiation and migration that determine what section of the developing embryo will become the head and which will become the skin, internal organs, and lower extremities. An opening in the embryo becomes what is called the "primitive streak," which allows the movement of cells within the embryo to form a set of layers. Cells that migrate to the "endodermal" layer begin the development of the gut and respiretory system, while cells forming the "mesodermal" layer begin the creation of the muscular and skeletal system. The cells moving to the "ectodermal" layer are of two types: one type forms the skin, while the other—the neuroectodermal stem cells (that are referred to more simply as neural progenitor cells)—develop into the brain and nervous system.

From the first divisions of the fertilized embryo, cells have different internal cellular environments as they divide. They therefore express different sets of genes and produce different sets of proteins encoded by those genes. This

---

[1] While there are many overviews of fetal brain development, I have relied in particular on the very clear account in Joan Stiles, Timothy T. Brown, Frank Haist, and Terry L. Jernigan, "Brain and Cognitive Development," in Richard M. Lerner, ed., *Handbook of Child Psychology and Developmental Science, 7th Edition*, (John Wiley & Sons, Inc., 2015)

sequenced differentiation controls how the embryo—and the brain within the embryo—develop. In the account that follows, there is much discussion of cells following chemical gradients both in migrating and in becoming particular types of cells. While these developments are driven by sequences of genes being expressed, the study of these processes belongs to the field of epigenetics, which explores the interplay of cellular chemical signaling and gene expression. The central point I would like to stress is that while the early development of the fetal brain is "genetically controlled," the work of this development is achieved through the complex mediations of cellular processes rather than some simple reading of instructions encoded in the genome.[2] The modes of mediation in this early phase of development—the nature of the biological information used in development—are profoundly different from the process of neuronal encoding that organize neural networks in the next developmental phase as the brain increasingly incorporates the "external" information through shaping the synapses and long-distance myelination.[3]

The neural progenitor cells form a structure called the neural tube. The tube itself gradually develops into the ventricular system of the brain, and the progenitor cells initially form a single layer surrounding the tube. This layer becomes the "ventricular zone" that will be crucial in the development of the brain. Cells at one end of the ventricular zone begin to differentiate into the cells for the forebrain (the "telencephalon" and the "diencephalon"), midbrain and hindbrain, while the other begins to form the spinal cord. The neural progenitor cells begin to proliferate through mitosis and to migrate away from the

---

[2] As Stiles et al., explain,

> What is inherited at conception is quite specific: (a) the DNA, and (b) the first cell with the cellular machinery for translating the information in the nucleotide sequences of DNA into proteins (the active agents in all biological processes). Biological inheritance provides essential tools, but neither the genes nor environmental factors prescribe outcomes. Rather brain development proceeds via the complex interaction of molecular, cellular, and environmental systems and elements.

Stiles et al., "Brain and Cognitive Development," p. 2

[3] For the brain, all information—interoceptive, proprioceptive, and exteroceptive—is external.

ventricular zone. The neural progenitor cells in the ventricular and subventricular zone (SZ) begin to undergo *asymmetrical* cell division in which the progenitor cells divide to produce one progenitor cell and one neuron. The transformation of the inchoate telencephalon into the neocortex thus begins with the migration of neurons away from the ventricular zone. These initial neurons form a layer called the preplate (PP in Figure 5.1), which, with the arrival of more neurons, further splits into a "marginal zone" (MZ) and a subplate (SP). The marginal zone and subplate serve as the scaffolding from which the fetal brain develops but then disappear through cell death before birth.

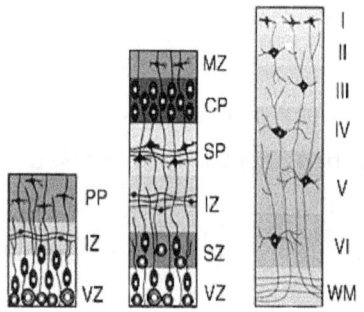

Figure 5.1: The Fetal Forming of the Layers of the Cortex[4]
PP= Preplate (splits into MZ and SP), IZ = Intermediate Zone (matures into white matter layer), VZ = Ventricular Zone, MZ = Marginal Zone, CP = Cortical Plate (matures into the cortical sheet), SP = Subplate, SZ = Subventricular Zone)

As yet more neurons migrate away from the ventricular zone, they position themselves in a new emerging layer—the cortical plate—between the marginal zone and the subplate. In the cortical plate, the neurons that arrive first become layer VI, the deepest layer of the developing neocortex, while those that arrive later move beyond the earlier neurons to form layers V, IV, III, II, and I in inverse order. Still, the neurons that create the cortical sheet do not develop uniformly, since the forebrain has an emerging internal structure that separates distinct regions for the motor cortex, the somatosensory cortices, and the visual cortex. This differentiation relies on a pair of gradients of signaling proteins produced at opposite ends of the ventricular zone. At the anterior end, the high concentration

---

[4] Stiles et al., "Brain and Cognitive Development," p. 12.

of one protein and the low concentration of the other leads to neurons differentiated for the motor cortex, while the opposite set of concentrations at the posterior end leads to neurons specific to the visual cortex. The neurons for somatosensory cortices begin to develop between the two ends.

The processes of development driven by the epigenetic unreeling of information encoded in the genes proceeds until all the major brain structures are formed.[5] In addition to the formation of these brain regions, however, the epigenetic sequencing also drives the development of neuronal connections both *between* regions and *to* the neurons from the spinal cord and the sensory organs (although in some cases, the establishment of synaptic connections begins to draw on external, experience-driven input). For example, Stiles et al. describe the corticospinal tract that connects the muscles of the body to the motor cortex:

> The CST [corticospinal tract] forms a conduit between the motor cortex and the limbs of the body, such that activity in the motor cortex depends on the motor activity of the limbs and vice versa. This activity drives the maturation of the CST....[6]

Similarly, the pathway from the retina through the thalamus to the visual cortex appears to rely in part on spontaneous firing of the fetal retinal ganglion cells.[7] Other major fetal connections, however, remain largely independent of the experiential stabilization of synapses. The crucial corticothalamic and thalamo-

---

[5] Stiles et al. provide a detailed account of the embryonic and fetal development of the brain that includes the importance of the glial structural cells that I leave out of this account to avoid overburdening the reader.

[6] Stiles et al., "Brain and Cognitive Development," p. 14.

[7] Kristin Keunen, Serena J. Counsell, and Manon J.N.L. Benders, "The emergence of functional architecture during early brain development," *NeuroImage* 160 (2017):

> Once thalamocortical connections are established, which has been noted to occur between 24–32 PCW (…), sensory stimuli including visual and auditory input can reach the developing cortex. The formation of transient thalamocortical-subplate circuits has been linked to the emergence of *spontaneous activity transients* (SATs) on electroencephalography recordings in extremely preterm infants (…). SATs are endogenous bursts of neuronal activity, which are either autonomously generated in the cortex, or in response to input from the subplate and are thought to drive neural circuit formation before sensory stimuli come online. (p. 4)

cortical pathways (CTP and TCP) that link the sensory cortices to the thalamus as the major relay site for activations from the sensory organs (the retinas and cochlea as well as the muscles and skin) develop through the mediation of neurons of the temporary subplate region. In the TCP, neurons from the thalamus extend axons to the subplate, and these neurons in the subplate extend axonal connections to the cortex. The subplate connections then guide the creation of the synaptic connections from the thalamus to the cortices. In the CTP, the process is reversed: neurons in the developing cortex send axons to neurons in the subplate, which then sends axons to the thalamus. This process is epigenetic rather than experience-driven. Similarly, the long-distance pathways which, after myelination, become the major white matter tracts all form during the prenatal period, even if the postnatal myelination is experience-driven.[8] The processes of neuron production and migration largely occurs prenatally, and almost all the neurons in the brain are created before birth. Indeed, recent research has increasingly stressed that even though the frontal lobe is very immature in neonates, the essential neuronal scaffolding is present and adequate to support developmentally important functional connectivity in the early months of life.[9]

The final major epigenetically triggered phase of development in the brain begins before birth with apoptosis, the programmed death of neurons in the subplate and ventricular zones, paired with a burst of synaptogenesis (the forming of synaptic connections), the arborization of neurons (the proliferation of "branches" on the axon) that produces a profusion of synaptic connections and contributes to a rapid growth in cortical surface area and thickness. However, the rate and timing of synaptogenesis are not uniform throughout the brain, and these differences in rate and timing have a profound effect on the process of

---

[8] See, for example, Douglas C. Dean III, Jonathan O'Muircheartaigh, Holly Dirks, Nicole Waskiewicz, Lindsay Walker, Ellen Doernberg, Irene Piryatinsky, and Sean C. L. Deoni, "Characterizing longitudinal white matter development during early childhood," *Brain Structure and Function* 220 (2015), pp. 1921-33.

[9] See, for example, Gal Raz and Rebecca Saxe, "Learning in Infancy Is Active, Endogenously Motivated, and Depends on the Prefrontal Cortices," *Annual Review of Developmental Psychology* 2020.2, pp. 247-68.

experience-driven development. In particular, while the lateral temporal, parietal, and prefrontal cortex expand in surface area "nearly twice as much as other regions" in the insular, medial temporal and occipital cortex, the high expansion occurs later than in low-expansion areas.[10] Jason Hill and his colleagues speculate on the evolutionary logic of this differential timing between high- and low-expanding regions:

> The preset result suggests that many cortical regions that expanded rapidly in evolution were also under evolutionary pressure to remain structurally immature during gestation. In particular, the lateral temporal, parietal, and frontal regions associated with high expansion in human postnatal development and in evolution are generally implicated in higher cognitive functions that distinguish humans from nonhuman primates.[11]

They suggest possible functions for the differences in timing in the process of experience-expectant development. In particular, the differential timing can "facilitate the contributions of postnatal experience to the development of selected regions...."[12]

While the timing and extent of dendritic arborization are epigenetically driven, the determination of which synaptic connections survive from all that are possible is a matter of experience-based activation rather than genetic coding. Here we return to the Hebbian learning rule that "neurons that fire together wire

---

[10] Jason Hill, Terrie Inder, Jeffrey Neil, Donna Dierker, John Harwell, and David Van Essen, "Similar patterns of cortical expansion during human development and evolution" *PNAS* 107.29 (July 20, 2010), p. 13138:

> In general, low-expanding regions tend to reach various structural and functional milestones earlier than do high-expanding regions. Mature synaptic density, peak cortical thickness, and mature values of gray matter density are reached earliest in low-expanding regions (V1 and Heschl's gyrus), later in intermediate-expanding regions (frontopolar and dorsal parietal cortex), and latest in high-expanding regions (DLPFC).

[11] Hill, et al., "Similar patterns of cortical expansion during human development and evolution," p. 13139.

[12] Hill, et al., "Similar patterns of cortical expansion during human development and evolution," p. 13139.

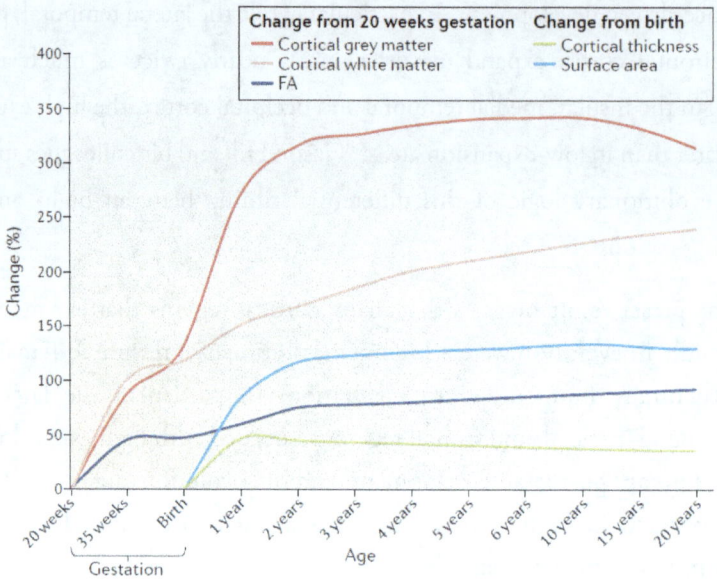

Figure 5.2: Developmental Timeline[13]
FA = Fractional Anisotropy, a measure of the maturation of white matter tracts.

together," while synaptic connections that are not activated die. As an additional factor in selection, the neurons compete for chemical signals—neurotrophic factors—that stabilize those neurons with active synaptic connections.[14] Thus the developing brain creates its systems of synaptic connections through early patterns of activation: Stiles et al. note, for example, that initially there are neurons whose synaptic connections link the retina to the primary *auditory*

---

[13] John H. Gilmore, Rebecca C. Knickmeyer and Wei Gao, "Imaging structural and functional brain development in early childhood," *Nature Reviews Neuroscience* 19 (March 2018), p. 127.

[14] Gregory Z Tau and Bradley S Peterson, "Normal Development of Brain Circuits" *Neuropsychopharmacology Reviews* (2010) 35, p. 152. Also see Stiles and Jernigan:

> Neurotrophic factors are produced by target neurons at synaptic sites, and are taken up by the afferent neurons that make effective connections with the targets (Huang and Reichardt 2001). During development it is thought that neurons compete for neurotrophic resources. According to the neurotrophic hypothesis (Oppenheim 1989), neurons that establish effective connections are able to obtain more neurotrophic factor and are more likely to survive.

Joan Stiles & Terry L. Jernigan, "The Basics of Brain Development," *Neuropsychology Review* 20 (2010), p. 339.

cortex but that these neurons die and disappear through the early competition to create viable active connections.[15]

A second crucial experientially driven process begins to stabilize the active synaptic connections and the emerging neural networks that the synaptic connections are starting to define. This second process is myelination. As discussed in Chapter Two, myelin is a mix of proteins and phospholipids that forms a sheath around the neuron's axon to greatly enhance the strength of the electrochemical spike of depolarization that serves as the activation signal transmitted from the cell body to the synapses of the axonal tip. The myelin sheath, however, comes from a type of glial cell called an oligodendrocyte. While oligodendrocytes begin to create myelin sheaths around neurons in some subcortical regions and in parts of the sensorimotor cortex prenatally, myelination increases rapidly—but selectively—after birth (see Figure 5,3.).

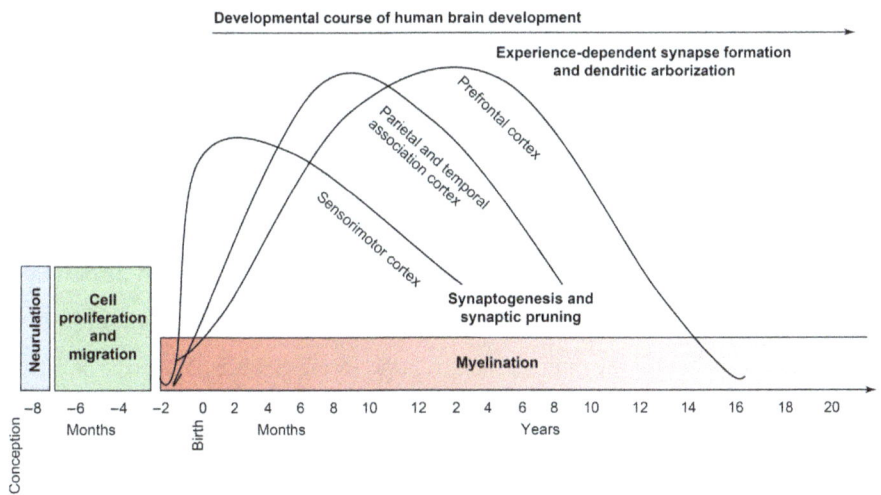

Figure 5.3: Neuronal Developmental Timeline[16]

---

[15] Stiles et al., "Brain and Cognitive Development," pp. 22-23.

[16] The figure is from Baars and Gage, *Cognition, Brain, and Consciousness*, p. 479, modified from R.A. Thompson and C. A. Nelson, "Developmental science and the media. Early brain development," *American Psychologist* 56 (2001), 5–15. Source: B. J. Casey et al., "Imaging the developing brain: What have we learned about cognitive development," *Trends in Cognitive Science* 9.3 (2005), pp. 104-110.

Oligodendrocytes appear to respond to neuronal activity in forming myelin sheaths at the same time that synapses are also stabilizing through repeated activation or are being pruned for lack of activation. Out of this process neuronal networks emerge. However, it is profoundly important that myelination proceeds from the back of the brain forward, with the sensory cortices first, then the association cortices, and then finally the prefrontal cortex (See Figure 3). Synaptogenesis, sequenced myelination, and synaptic pruning shape the period of *experience-expectant* postnatal development:

> Experience-expectant processes rely on the developing organism being exposed to basic environmental information (e.g., patterned light, sequential sounds), so rudimentary that it should be universally experienced by all members of a species across variable environments. Experience-expectant processes are linked to sensitive periods: windows of time in which a developing system is highly plastic and most open to influence by the environment (e.g., Hensch, 2005). Tuning of the brain to the expectable environment during sensitive periods is manifested as retention of a subset of necessary synapses among those that were initially overproduced (Greenough et al., 1987), reducing the amount of brain development that must be precisely specified by genetics.[17]

The order of myelination from the sensorimotor cortices to the association cortices to the prefrontal cortex also determines the order of stabilization of local neural networks before they participate in larger-scale systems. Sensitive periods, like myelination, thus appear in a back-to-front sequence. As Eric I. Knudsen explains:

> A sensitive period cannot open until three conditions are met. First, the information provided to the circuit must be sufficiently reliable and precise to allow the circuit to carry out its function (for high-level circuits, this may not happen until relatively late in development). Second, the circuit must contain adequate connectivity, including

---

[17] Amanda S. Hodel, "Rapid infant prefrontal cortex development and sensitivity to early environmental experience," *Developmental Review* 48 (2018), pp. 120-21.

both excitatory and inhibitory connections, to process the information (Fagiolini & Hensch, 2000). Finally, it must have activated mechanisms that enable plasticity, such as the capacity for altering axonal or dendritic morphologies, for making or eliminating synapses, or for changing the strengths of synaptic connections.[18]

Information reliability depends on myelination of the afferent (i.e., input) neurons providing data as well as the stabilization of local networks. And indeed, it turns out that the input streams for the sensory cortices are myelinated prenatally so that, with both synaptogenesis and synaptic pruning actively shaping the neural networks, the sensory cortices transition to sensitive periods beginning at birth.[19] In contrast, the association cortices, which receive their bottom-up input from the sensory cortices (as well as subcortical nuclei), must wait for the networks of the sensory cortices to be sufficiently stable and precisely tuned before they can begin their sensitive periods.

Chapter 3 discussed the way in which in the visual system, higher-level association areas in the parietal cortex create increasingly higher dimensional spaces to represent mutually differentiated distributions of ever more complicated objects. In subsequent processing, the temporal and frontal cortices combine the visual information with information from other subcortical and cortical networks to produce multimodal representations both in perception and in memory. The mathematical models in Chapter 1 described ways in which these hierarchies of networks could be constructed. The developmental model of a sequence of sensitive periods for brain regions that move from back to front and that derives from progressive myelination, prolific synaptogenesis, and

---

[18] Eric I. Knudsen, "Sensitive Periods in the Development of the Brain and Behavior," *Journal of Cognitive Neuroscience* 16.8 (October 2004), p. 1414. This is perhaps the classic article on the topic.

[19] Gregory Z Tau and Bradley S Peterson note:

> Between GA [gestational] weeks 20 and 28, mature myelin is detected first in subcortical regions and later in cortical regions. At GA week 35, it is detected in the precentral and postcentral gyri and optic radiation, and at GA week 40, it is present in the acoustic radiation (Iai *et al*, 1997).

Tau and Peterson "Normal Development of Brain Circuits," p. 35.

experience-driven synaptic pruning now provides the mechanism by which these hierarchies of networks come into being. Also, note that, for any given region, because the higher-level networks have not yet formed, higher-level networks cannot provide coherent, stable top-down feedback to participate in the initial shaping of the neuronal networks during the region's sensitive period. In contrast, it is important, as will be discussed below, that the subcortical nuclei provide additional input to the developing neuronal networks and have a powerful effect in shaping the emerging representational spaces.

Getting it Together: Core Functional Networks

Much research remains to be done to clarify the developmental trajectories of the various cortical regions. In part, this task presents a technical challenge: there are strict limits to the sorts of invasive methodologies that can be applied to human infants, and imaging studies in infants are difficult because babies move around. Still, because neuroscientists increasingly understand brain activity as the interaction of networks across broad areas rather than simply as isolated patterns of local activation, developmental neuroscientists have turned their attention to the slow emergence of the core networks through which humans engage the world and their own selves. These growing networks include the attentional and salience networks discussed in Chapter 2 as part of the visual system as well as the default mode network (DMN) and the systems for episodic and semantic memory that feed into the human networks supporting language. Another central network whose development researchers are attempting to trace is that of the social self. The "self" system is, at the same time, a "social" system since it is within the initial dyadic relation with a caregiver that an infant begins to plan and produce the sorts of voluntary actions that define the self as an agent in the world. In the following section, I discuss research on the more general networks like those for salience, attention, and the default mode network. The beginnings of a "self" system will be the topic of the next section because it appears very early during infancy.

The explorations of vision and emotion in Chapter 3 and 4 have shown not simply that the brain works through a complex of neuronal networks with bottom-up, top-down, and lateral components, but that the experiences shaped by all of these interactions integrate a vast array of internal resources, from the sensory data, to memories, to plans, to one's current emotional state. These patterns of interaction do not so much mediate experience as define it. Thus, a key developmental question is how the networks that construct these interactions arise. In probing the emergence of networks, researchers have drawn on graph theory—most commonly known through social network analysis (SNA)—to make sense of the imaging data, just as neuroscientists looking at *local* network structures and dynamics rely on models from mathematical neural network modeling to understand relations between synaptic organization and function at the cellular level. Network neuroscientists have established a few general patterns for the network organization in the neonate brain that provide a base from which to explore developmental changes. Since their research is based on imaging, which has limited resolution, the *nodes* in the networks —the units connected together to make the network—are active *clusters* of neurons (rather than discrete neurons), and the *edges*—the connections between units—are not the synaptic connections between individual neurons but are the *average synaptic connections* of those clusters of neurons. Kristen Keunen lists a set of significant graph-theoretic properties that appear in the neonate neuronal networks. First, the *characteristic path lengths* (the distance between connected nodes) are short, and the nodes tend to be connected to their neighbors (i.e., *clustered*). Second, given these two properties (short path lengths and clustering), the networks reveal *small world* organization.[20] Next, the networks are *modular* rather than highly integrated. And finally, the total infant network shows a *rich man's club* organization: the nodes with the highest number of connections are all connected to one another.[21] Starting from these early modular networks, developmental change involves the slow integration of the networks *across* regions, accompanied by a simplification of local networks *within* a region, as aspects of their functions shift to the more efficient and effective *long-range networks*.

---

[20] There are many discussions of the properties of "small world" networks. See, for example, Duncan J. Watts, *Six Degrees: The Science of a Connected Age* (New York: W.W. Norton and Company, 2003), pp. 69-100.

[21] Kristin Keunen, et al., "The emergence of functional architecture during early brain development," p. 5.

The infant system for the control of attention is one category of networks that has long been of interest to developmental psychologists since well before the advent of the new imaging technologies. For the first few months, a neonate cannot control what it looks at, but soon it develops enough voluntary control of attention to turn away from objects dangled before it.[22] Michael Posner and Mary Rothbart have provided a framework of three systems through which to formulate the development of attentional control that extends beyond the association cortices to slowly emerging frontal networks:

> The alerting network is related to the functions of obtaining and maintaining the alert state. The orienting network is involved in the selection of sensory events. The executive network is involved in resolving conflict among response tendencies.[23]

Each of these networks relies on a different neuromodulator and on networks in different regions:

> The *alerting network* is modulated by the brain's norepinepherine system arising in the midbrain and making contact with frontal and parietal areas. The *orienting network* involves areas of the inferior and superior parietal lobe and the frontal eye fields. Cholinergic systems arising in the basal forebrain play a critical role in modulating the orienting network.... The *executive network* involves the anterior

---

[22] In more technical language:

> Neonates exhibit complex, self-generated, but disorganized movements that do not seem goal oriented (...). Eye saccades at this age are elicited reflexively and are directed by environmental stimuli rather than through endogenous control (...), consistent with maturation of somatosensory and motor areas ahead of the visual and association cortices. The behavioral changes that mark the third postnatal month, including inhibitory control over reflexive behaviors and saccades, as well as goal-directed behaviors such as target-directed head–eye coordination and reaching to grasp, may reflect the cortical remodeling and myelination of association areas occurring at that time (...).

Tau and Peterson, "Normal Development of Brain Circuits," p. 156.

[23] Michael I. Posner, Mary K. Rothbart, Brad E. Sheese, and Pascale Voelker, "Control Networks and Neuromodulators of Early Development" *Developmental Psychology* 48.3 (May 2012), p. 828.

cingulate gyrus, anterior insula, basal ganglia and parts of the prefrontal cortex. These areas are rich in dopamine and their function is modulated by dopamine from the ventral tegmental areas.[24]

These different systems, based on different neuronal networks, mature at different times:

> [E]ven newborn babies have the capacity for alerting, in its most basic form. However, the more-complex visual attention-orienting mechanism, which allows for suppression of competing information during attention-orienting shifts, becomes functional only between 4 and 6 months of age. Before this age, attention orienting in infants primarily consists of simpler processes that facilitate the orienting of the infants' attention towards perceptually salient information.[25]

The executive network, involving the most complex integration of information, matures even later in childhood and extends into early adulthood. Since theorists of predictive processing in the brain have described visual attention as a crucial mechanism for improving the quality of the visual data needed for accurate prediction, other developmental issues are tied to increasing control of visual attention. In particular, Amso and Scerif suggest a virtuous cycle in the visual system that begins with the initial phase of increased attention control:

> [I]n early postnatal life, vision is poor and feedforward visual information conveyed to higher cortical areas is minimal. We hypothesize that, with visual development, there is an increase in feedforward information competing for attention allocation in higher-level regions, thus linking top-down visual attention development with visual experience. In turn, these regions, now engaged, send top-down signals to begin to tune local visual areas,

---

[24] Posner et al., "Control Networks and Neuromodulators of Early Development," pp 828-29.

[25] Dima Amso and Gaia Scerif, "The attentive brain: insights from developmental cognitive neuroscience," *Nature Reviews Neuroscience* 15 (October 2015), p. 607.

setting the hierarchical loops in motion from very early in the first postnatal year.[26]

The increased functioning of higher-level association areas drives the ability to discriminate and learn about objects (and to link experiential data from other modalities to these objects). Thus, tracing the addition of an orienting network to the salience-driven, largely automatic alerting system at the same time as the association cortices are undergoing a sensitive period greatly increases the power of perceptual processing and learning.

While the maturation of the association networks of the parietal and temporal cortices is crucial in the early development of attentional control, the attentional network, as in adults, also involves the lateral prefrontal cortex which provides the capacity for higher-order integration of sensory and interoceptive data with nascent memory structures and subcortical evaluative systems. It turns out that by the third month postnatal, the infant brain begins to incorporate the prefrontal cortex—despite its structural immaturity—into its attention and salience networks.[27]

A second network—the default mode network (DMN)—usually is treated as complementary to the attentional networks: when the attentional networks become active in assessing the significance of interoceptive, proprioceptive, and exteroceptive data, the DMN, which monitors internal states, goes quiet. Since the DMN is associated with the sense of a "self," researchers have been very interested in the timing and manner of its emergence.[28] This new focus

---

[26] Amso and Scerif, "The attentive brain," p. 609.

[27] Raz and Saxe in "Learning in Infancy Is Active, Endogenously Motivated, and Depends on the Prefrontal Cortices" acknowledge that cortical connections to the prefrontal cortex are not yet myelinated, and there is little of the robust synaptic arborization and experience-dependent pruning that comes later, but they suggest that, for 3-month old infants, connectivity to the sensory and association cortices may be mediated by already fully myelinated corticothalamic connections among the regions.

[28] For example, see Chiara Bulgarelli, Anna Blasi, Carina C.J.M. de Klerk, John E. Richards, Antonia Hamilton, and Victoria Southgate, "Fronto-temporoparietal connectivity and self-awareness in 18-month-olds: A resting state fNIRS study," *Developmental Cognitive Neuroscience* 38 (2019):1-12

reflects a broader shift in thinking about the maturation of functions in the brain. While one ongoing project in neuroscience is to clarify the functions of specific brain regions (dorsolateral as opposed to ventromedial prefrontal cortex, etc.), network neuroscience has shown the centrality of the integration of regions into networks, an integration in which specific regions can serve as nodes in several *different* networks. The challenge of developmental network neuroscience, is to understand the processes by which the various different key networks are assembled. In particular, because of the back-to-front sequencing of synaptogenesis and myelination, the DMN, dependent on networks in the prefrontal cortex, develops slowly and late compared with the subcortical, sensory, and motor networks. Josepheen De Asis-Cruz and her group, for example, identify four modules ("nodes that share dense connections with nodes within their group and sparse connections outside their group") in the neonate neuronal network: (1) a somatosensory/motor subsystem, (2) an occipital module, (3) a fronto-temporal module, and (4) a limbic-paralimbic-subcortical module. These discrete modules suggest that, while components of the adult DMN have begun to appear, a functioning DMN has not yet developed.[29] Recent research like that of

---

[O]ur current knowledge of early self-awareness is limited, while much work in adult cognitive neuroscience has already made significant progress in identifying the neural underpinnings of self-related processing. Specifically, a network of brain regions which is activated during passive rest (i.e. resting-state) in the low-frequency range (< 0.1 Hz), appears to be recruited during self-related processing (Raichle, 2015). This so-called Default Mode Network (DMN), which overlaps considerably with the social brain network (Mars et al., 2012), is composed of the medial prefrontal cortex (mPFC), the precuneus, the posterior and anterior cingulate cortex, the inferior parietal lobe (IPL), the medial temporal lobe and the temporoparietal junction (TPJ). (p. 2)

[29] Josepheen De Asis-Cruz, Marine Bouyssi-Kobar, Iordanis Evangelou, Gilbert Vezina, and Catherine Limperopoulos, "Functional properties of resting state networks in healthy full-term newborns," *Nature: Scientific Reports* (December 2015) DOI: 10.1038/srep17755:1-15.

However, to complicate the story a bit, the most recent analyses of the *adult* default mode network have increasingly concluded that, rather than a single network, it is comprised of at least two parallel sub-networks, one that handles the integration of autobiographical memories into assessments of current circumstances and a second that assesses the inner states of others (theory of mind.) Randy L. Buckner and Lauren M. DiNicola, "The brain's default network: updated anatomy, physiology and evolving insights," *Nature Reviews Neuroscience* 20 (October 2019), p. 600. The authors note, "These features [of the DMN]— expansion,

Amanda Hodel continues to stress the late development of the prefrontal regions needed to support the DMN. Still, Hodel, like Raz and Saxe, argues that most of the basic work in defining the connectivity of the DMN to regions in the prefrontal cortex appears to be completed (if not fully myelinated) by the end of the first year after birth.[30] The maturation of the DMN largely completes the arc of development from the epigenetically shaped structures of the fetal brain to the experience-expectant connection of synapses, regions, and networks. But depiction of the DMN as the "self" network raises a question.[31] How does a self emerge from the maturation of all of these networks? To begin to answer this question requires a shift in focus from general consideration of developmental processes to the role of subcortical structures in shaping the developing brain.

The "Self" in the System

THE ROOTS OF ACTION IN PRIMARY SENSORIMOTOR INTENTIONALITY

In 1966, Carolyn Rovee-Collier, a graduate student writing her thesis in child psychology, invented the expedient of using a string to connect a mobile to the leg her 3-month-old son to try to keep him amused.[32] She realized that not only had he, to his delight, discovered the correlation between kicking his leg and the mobile moving, but he also remembered this correlation. It took persistence for her to persuade her peers of this ability—which contradicted contemporary theories about infant memory—but this now much-replicated experiment has become a classic in the field. I will return to the problem of memory in the next

---

differentiation and freedom from the strong constraints of sensory hierarchies — might have been critical evolutionary gateways accounting for their functional roles in humans."

[30] Hodel, "Rapid infant prefrontal cortex development and sensitivity to early environmental experience" p. 119.

[31] For an excellent account of current approaches to the DMN, see Yaara Yeshurun, Mai Nguyen and Uri Hassan, "The default mode network: where the idiosyncratic self meets the shared social world," *Nature Reviews Neuroscience* 22 (March 2021), pp. 181-92.

[32] The wonderfully dry title of the paper reporting her replication of this experiment with a larger group of infants is "Conjugate Reinforcement of Infant Exploratory Behavior" (*Journal of Experimental Child Psychology* 8 [1969], 33-39). See her obituary in the *New York Times*, October 22, 2014.

chapter and, for the moment, turn to the question of why her son was amused. In proposing an answer, I draw on the elegant reconceptualization of Jaak Panksepp's model of primary process emotions offered by Jonathan T. Delafield-Butt and Nivedita Gangopadhyay.[33] They center their account on the concept of sensorimotor intentionality, adopted from the phenomenological philosophy of mind and, in particular, on

> a primary sensorimotor intentionality that develops in cognitive sophistication and motor precision during ontogenesis, but remains fundamentally unchanged and continuous throughout life, driving and shaping development, learning, and cognition.

"Intentionality," as they explain, is the "being-about" quality of a mental state (including perceptions, emotions, and memory), and they argue that brain activity from the very beginning is "about" something. Their starting point is the idea that brains are for action, which for humans introduces the idea of agency: that one (a self) acts in the world.[34] What binds brain activity to being "about something" is the nexus of brainstem nuclei that are functional from before birth:

> Brain stem territories include the necessary functional characteristics for enabling a primary form of consciousness as 'acting with knowing' – an embodied and prospective agentive experience. Somatotopic mapping giving access to *proprioceptive* body-space and tactile information is preserved in the brain stem…, distance receptor projections from eyes, ears, and nose laminate in the roof of midbrain giving *exteroceptive* sensory information, and visceral organ function

---

[33] Jonathan T. Delafield-Butt and Nivedita Gangopadhyay, "Sensorimotor intentionality: The origins of intentionality in prospective agent action," *Developmental Review* 33 (2013) 399–425.

[34] Delafield-Butt and Gangopadhyay, "Sensorimotor intentionality," p. 402. They cite Roger Sperry's polemical assertion that "the sole product of brain function is motor coordination" [Roger W. Sperry, "Neurology and the mind–brain problem." *American Scientist*, 40 (1952), 291–312.], but also draw connections to the work on embodied and enactive cognition by Andy Clark, Evan Thompson, and Alva Noë.

is monitored, especially in periaqueductal grey, giving *visceroceptive* information on the body's vital wellbeing and physiological need.[35]

The brainstem and midbrain nuclei provide the fundamental motivating assessment of the world and the body that directs agency. Delafield-Butt and Gangopadhyay stress, however, that while these nuclei provide the primary form of agentive action that is necessarily operative at birth if the infant is to survive, the cortical elaboration of agency and action creates the flexibility and responsiveness to complex circumstances that characterize human agency:

> As development proceeds, cortical regions become increasingly important contributors giving new cognitive capacities to the agent for 'abstraction' or disengaging from immediately perceived environmental affordances and engaging with (new) cortically-generated ones, *c.f.* 'objects', by forming beliefs, plans, and conceptual understanding of the world. Cortex together with central limbic structures enable expanded action plans that draw on previous experience. The more advanced the level of development and cortical maturation, the more integration of different brain processes becomes involved (Vandekerckhove & Panksepp, 2011).[36]

The connection here with Panksepp's work is crucial because the subcortical nuclei that serve as the epigenetically shaped primary intentional matrix are not neutral, abstract generalities but have very specific features that have emerged through evolutionary pressures. These nuclei—responding to the earliest forms of interoceptive, proprioceptive, and exteroceptive information—generate the assessments and impulses toward action of Panksepp's SEEKING, LUST, PLAY, and other systems.

Although Panksepp's and other researchers' work on animal systems have articulated some of the behavior of these early brainstem and midbrain networks, they largely remain something of a black box for humans. At present we can chart some of the functions of the neuromodulatory systems and their behavioral

---

[35] Delafield-Butt and Gangopadhyay, "Sensorimotor intentionality," p. 408.

[36] Delafield-Butt and Gangopadhyay, "Sensorimotor intentionality," p. 409.

consequences for development—with attachment as a particularly important case to be discussed below—but even before we have a detailed understanding of the systems, Delafield-Butt and Gangopadhyay's model allows us to think through the larger implication of the brainstem as a genetically shaped source of primary sensorimotor intentionality. Delafield-Butt and Gangopadhyay primarily are concerned to provide a model of embodied mind and cognition that replaces Cartesian dualism—a concern I certainly share—but I also offer this model in order to greatly complicate the emphases on homeostasis and allostasis that appear in the predictive coding models for human action.[37] Carolyn Rovee-Collier's son joyfully learned to move the mobile with his foot with an attention to action shaped by the congenital SEEKING and PLAY systems in which calculations based on allostasis had no meaningful role.

Taking the nuclei in the brainstem as the genetically shaped source of primary sensorimotor intentionality repeats at a system level the sort of logic seen in the development of the brain at the level of synapses, regions, and transregional connectivity. The early maturation of the subcortical nuclei allows them to serve a bootstrapping function for neonate action, and those nuclei are ready at birth to mediate the experientially driven articulation of neuronal networks. They make the remembered experiences "about" something to the inchoate self.

The Emergence of the "Social" in Early Development

The next phase of experience-driven shift from subcortical control to cortically centered networks for action during the first year after birth is a complex story with a remarkable result: the emergence of a "social" self system. However, it is

---

[37] They explain that

> our proposal of grounding development of cognition in primary sensorimotor intentionality enables an account of the mind and nature of cognitive mechanisms underlying mental states as necessarily grounded in sensorimotor goals and their spatiotemporal frames of reference, rather than as described within a Cartesian mechanistic framework. Intentionality or the object-directed nature of mental states is secured by the successful continuation of the structure of sensorimotor intentionality in higher cognitive processes. Sensorimotor intentions give the framework of an embodied mind, from simple motor activity to abstract thought. (p. 411)

crucial from the outset to see this social "self" as part of a yet larger system for the shaping of action. Indeed, the early developers of attachment theory—which tracks the processes of forming social bonds—insisted that attachment (and the social realm) is but one of a set of four interrelated behavioral systems: attachment, wariness, exploration, and sociability.[38] Thus the account that follows centers on the emergence of the social self traced in attachment theory but also of necessity investigates the roles of fear, exploring, and sociability in shaping this self.[39]

This story weaves together accounts from current neuroscientific research, from the many years of careful observation both by developmental psychologists and by researchers in attachment theory in particular. The three quite separate disciplines tell essentially the same story—largely agreeing to the same set of facts—but with different interests. Their respective story lines increasingly converge as noninvasive imaging techniques have allowed researchers to link behavioral transformations to cortical developments. Much work in developmental neuroscience has been devoted to studying the gradual maturation and integration of neuronal networks and the evolving network features of early neuronal systems, but as knowledge of these developments grows, researchers are asking questions about the emergent behavioral properties generated by these systems. Developmental psychology for decades has tracked the stage-by-stage growth of *behavioral* abilities from the simplest tasks of eye movement and grasping to the development of semantic memory, to the understanding of the intentions of observed actions, producing a large body of

---

[38] See, for example, the discussion in Robert S. Marvin, Preston A. Britner, and Beth S. Russell, "Normative Development – The Ontogeny of Attachment in Childhood," in *Handbook of Attachment: Theory, Research, and Clinical Applications*, edited by Jude Cassidy and Phillip R. Sharver, 3rd edition (New York: The Guilford Press, 2016), pp. 275-76.

[39] Jaak Panksepp included PLAY in his list of primary affective systems, and comparative developmental psychologists increasingly consider play—as the early-manifesting behavioral system for sociability—to have a an important developmental role:

> Hinde (1974) described nonhuman primates' play with peers, which he identified as different from mother-child interactions, as "consume[ing] so much time and energy that it must be of crucial adaptive importance."

Jude Cassidy, "The Nature of the Child's Ties" in Cassidy et al., *Handbook of Attachment*, p. 9.

external observational data that now can be correlated with the development of neuronal networks that are tracked by neuroscience. And similarly, while attachment research historically has been about attachment *behavior*, it increasingly has sought to integrate studies of the neuronal networks that produce the emergent observable behavior.

I have discussed the development of the brain from conception through the first year of life, but what in fact can a neonate do with the systems that are in place at birth? From the neuroscientific perspective, the gross structural features—including the long-distance connections that, once myelinated, become the white matter tracts—already are present. But developing the internal networks to process sensory as well as interoceptive data awaits the input that will shape synaptic connections and myelination. Hence, while the brain is in fact very active, neonates can *do* very little:

> Neonates exhibit complex, self-generated, but disorganized movements that do not seem goal oriented (…). Eye saccades at this age are elicited reflexively and are directed by environmental stimuli rather than through endogenous control (…), consistent with maturation of somatosensory and motor areas ahead of the visual and association cortices. The behavioral changes that mark the third postnatal month, including inhibitory control over reflexive behaviors and saccades, as well as goal-directed behaviors such as target-directed head–eye coordination and reaching to grasp, may reflect the cortical remodeling and myelination of association areas occurring at that time.[40]

From the developmental perspective, neonates before the age of 4 months have little control over where they look. They cannot direct their orienting or the attention that guides orienting.[41] Yet, despite their limitations, neonates do look

---

[40] Gregory Z Tau and Bradley S Peterson, "Normal Development of Brain Circuits" *Neuropsychopharmacology Reviews* 35 (2010), p. 156.

[41] Mary K. Rothbart, Brad E. Sheese, M. Rosario Rueda, and Michael I. Posner, "Developing Mechanisms of Self-Regulation in Early Life," *Emotion Review* 3.2 (April 2011), p. 208.

Researchers in attachment bring a slightly different focus and concern: in addition to looking, they stress the importance of visually guided grasping, which is the most basic form

around, and the looking has distinctive patterns. Starting as neonates, infants prefer to look at faces or objects that look like faces, and they prefer their caregiver's face in particular.[42] However, faces are in fact complex visual objects, and how a neonate looks at faces shifts over the first few months:

> Very young infants (i.e., under two months) tend to direct attention to the outer regions of the face and head, but by three to four months of age, children direct a majority of their visual attention to the eye region. In complex displays, however, faces do not capture the attention of children until around six months of age....[43]

Faces not only have an intrinsic interest, but they acquire valences—even if only within the category of familiar/unfamiliar. By six months, infants learn to recognize familiar faces seen at different angles and showing different emotional expressions.[44] Efforts have been made to explain how the trick is done. What prompts a neonate to look at faces and single out a caregiver's face in particular,

---

of attachment behavior but which also requires rudimentary control of both looking and reaching:

> Reaching, grasping, and clinging are also crucial attachment behaviors in all primates, and they develop relatively late in humans. It is not until about 2 months of age that the human infant's grasp is highly developed and controlled by anything other than a reflex-like process of activation by stimulation of the palm of the hand. It is at about the same time that the visual system becomes chain-linked with the motor system in a manner that allows the infant to make ballistic-like movement toward an object in the visual field.

Robert S. Marvin, Preston A. Britner, and Beth S. Russell, "Normative Development – The Ontogeny of Attachment in Childhood," in Cassidy and Sharver, eds., *Handbook of Attachment: Theory, Research, and Clinical Application*, p. 278.

[42] Laura J. Sherman, Katherine Rice, and Jude Cassidy, "Infant capacities related to building internal working models of attachment figures: A theoretical and empirical review," *Developmental Review* 37 (2015), p. 115. There is in fact much information on infants and face recognition in the developmental literature.

[43] Sherman et al., "Infant capacities related to building internal working models of attachment figures," p. 116.

[44] Sherman et al., "Infant capacities related to building internal working models of attachment figures," p. 116. As discussed below, amygdala function is depressed in very young infants so that the range of valences associated with objects being learned in their semantic memory structures is very narrow.

and why does familiar/unfamiliar matter?[45] One can point to interactions between the dopamine system and the release of oxytocin, and surely these play a central role. However, for the moment it is best, I think, to treat the subcortical networks behind primary sensorimotor intentionality—which as a general concept gathers together Panksepp's seven categories of primary affective dispositions—as a black box that simply somehow produces the inclinations observed in very young infants.[46] Even if the exact interactions that produce the neonate's behavioral propensities are not understood, researchers agree that these early-manifesting inclinations are evolutionary adaptations in humans that add biasing valences to the assessment of information (exteroceptive, proprioceptive, and interoceptive) at the time when the cortical systems for processing this information are being shaped. Overwhelmingly, the result of these biases is the creation of the infant as a social being. Ruth Feldman concludes:

> Humans are wired for social affiliation via activity of this limbic circuit, comprising the OT-producing hypothalamus, extended amygdala network, and striatum [including the ventral tegmental area (VTA), which projects to striatum, and VP, which receives projections from it].

Raz and Saxe stress:

> Infants are born into a social world. Humans are obligately social, depending on cooperation with others for food provision, childcare,

---

[45] Scholars who study attachment behavior are generally careful to use the neutral term "caregiver" rather than biological mother, since the crucial element is the role the primary caregiver plays in the initial dyadic relationship with the infant rather than the genetic connection itself.

[46] Ruth Feldman works on the neuroscience of attachment and focuses in particular on the interactions among neuromodulator systems and developing cortical networks across a range of mammals. Still, her research does not yet account for the shaping of human neonate visual attention. See, for example, Ruth Feldman, "The Neurobiology of Human Attachments," *Trends in Cognitive Sciences* 21.2 (February 2017):80-99. A recent study of the role of the frontal cortex in visual attention in early infancy simply concludes that "young infants pursue multiple, distinct intrinsic motivations" without further delineating how these motivations arise out of the subcortical networks. (Gal Raz and Rebecca Saxe, "Learning in Infancy Is Active, Endogenously Motivated, and Depends on the Prefrontal Cortices,.)

safety, and more. Human infants are hyperaltricial and thus even more radically dependent on other people to provide for their every need. In this context, infant behavior appears to be guided by a distinct motivation to form and sustain positive social relationships.[47]

Raz and Saxe make an additional point that informs the discussion below of an infant's increasing ability to take *active* roles in its engagement with its caregiver and the world. The early biases in attention form a feedback system in cortical development: "infants actively construct a curriculum for their own learning."[48]

Attachment

Infants during the first six months learn many things—from the coordination of sight and hand movement to the phonemes of spoken language—but in their constrained environment, they mostly learn patterns of human interaction. Quite remarkably, long before the rise of predictive processing as a paradigm in neuroscience, the idea of an infant's assimilation of models of behavior as a means to make predictions about the world and the self—and thereby to internalize and thus acquire modes of self-control modelled on caregiver action—was at the center of John Bowlby's initial formulation of attachment theory. Beginning in 1969, Bowlby proposed the idea of an inner working model (IWM) both of the caregiver whose behavior is to be predicted and internalized and of the infant itself in its participation in the evolving relationship with the caregiver. While Bowlby did not develop his account of the IWM in detail, this lack of detail actually seems to have had positive consequences: it left room for more recent neuroscientific ideas of predictive processes and interoceptive inference to construct a productive scaffolding that in turn allowed developmental and social neuroscience to assimilate the models offered by attachment theory into their own frameworks.

---

[47] Raz and Saxe, "Learning in Infancy Is Active, Endogenously Motivated, and Depends on the Prefrontal Cortices," p. 255.

[48] Raz and Saxe, "Learning in Infancy Is Active, Endogenously Motivated, and Depends on the Prefrontal Cortices," pp. 259-60.

One useful recent account of attachment and inner working models begins with basic inborn types of "fixed action patterns" that serve as forms of attachment behavior, most notably, grasping, crying, and smiling.[49] The earliest IWM is simply an internal representation of any predictable response dynamics to these initially fixed actions. If the caregiver is responsive to the infant's behavior, the infant's internalization of the dyadic patterns become increasingly stable, so that crying in particular as an attachment behavior begins to shift to other forms like looking at the caregiver and smiling.[50] Marvin et al. stress that these early interactions are repeated many times. It turns out that such repetition is important because, while the neonate hippocampus is still immature, it is nonetheless adequate to support the sort of statistical learning needed to create memory networks to link into the networks for perception, attention, and action-execution.[51] Once this initial network of networks is formed, it allows the infant to build both more complex response scenarios and a more complex IWM.

> First, during Phase II, there is an elaboration of simple behavioral systems into more complex ones. The simple behavioral systems of the Phase I infant become integrated within the infant into complex, chain-like behavioral systems. The primary focus here is on the control of the individual systems. Whereas in Phase I, the caregiver provides the conditions for terminating one behavioral link in a chain and activating the next, during Phase II, the infant assumes much of the control....
>
> A second defining issue for Phase II is the restriction of range of effective activating and terminating conditions.... Specifically, Phase II is operationally defined in terms of the infant differentiating

---

[49] Robert S. Marvin, Preston A. Britner, and Beth S. Russell, "Normative Development – The Ontogeny of Attachment in Childhood" in Cassidy et al., eds., *Handbook of Attachment*, p. 276.

[50] Marvin et al., "Normative Development – The Ontogeny of Attachment in Childhood," pp. 278-79.

[51] See, for example, Joan Stiles et al., "Brain and Cognitive Development," and Cameron T. Ellis, Lena J. Skalaban, Tristan S. Yates, Vikranth R. Bejjanki, Natalia I. Córdova, Nicholas B. Turk-Browne, "Evidence of hippocampal learning in human infants," *Current Biology* 31 (Aug. 2021):1-7.

between the most familiar caregivers and others in directing his or her attachment behavior....

A third and equally important component of Phase II is the infant's increasing tendency to initiate attachment-caregiving and sociable interactions with the principal caregiver(s). Ainsworth (1967) observed that as early as 2 months of age, and increasingly thereafter, infants are active in seeking interaction rather than passively responding to it....

Finally, these characteristics of Phase II have implications for describing the nature of the infant's IWMs. Most importantly, the infant can increasingly differentiate his or her primary caregiver(s) from others, and in that sense "know" who they are. However, the infant cannot yet conceive of the attachment figure as someone with a separate existence from his or her own experience.

While this account shows that one key concern among the researchers who study attachment is in the dyadic relation between the infant and the caregiver, an equally crucial question is how the dyadic relation opens up and enables the infant's exploration of the world. That is, from the beginning, attachment (and the constructions of IWMs) is not an end in itself but serves a creaturely function of acquiring modes of response to the world, including curiosity, fear, and broader social engagement.[52] Mary Ainsworth, following Bowlby, particularly pointed to the relationship between attachment and exploration:

> According to Ainsworth (1972), "the dynamic equilibrium between these two behavioral systems [attachment and exploration] is even more significant for development (and for survival) than either in isolation." [...] On the basis of her observations during the infant's

---

[52] Jude Cassidy stresses:

> The attachment behavioral system can be fully understood only in terms of its complex interplay with other biologically based behavioral systems. Bowlby highlighted two of these as being particularly related to the attachment system in young children: the exploratory behavioral system and the fear behavioral system....

Jude Cassidy, "The Nature of the Child's Ties," in *Handbook*, p. 8

first year of life, Ainsworth referred to an "attachment-exploration balance."[53]

Simply put, if an infant has successfully modeled the interactions with a caregiver that shape action and—increasingly—agency, this internal framework creates a "secure base" from which to explore the world.

Perhaps the clearest illustration of the shift from the dyadic relationship to one that increasingly encompasses the world is the process of developing a shared gaze, known as *joint attention*:

> As early as 3 to 5 months of age, attention coordination expands beyond the social dyad to include episodes of joint attention that describe triadic interactions between self, other, and an external object, event, or symbol.... The parallel processing of information about one's own and another's attention allows an infant to build up and internalize these representations and is considered essential for human learning. Between 3 and 9 months of age, the proportion of time infants spend in episodes of shared dyadic attention decreases substantially while the proportion of time spent in shared attention to objects increases rapidly.[54]

The infant's initial "response to joint attention" (RJA), like the earlier focus on faces, appears to be a form of first-order behavior built upon the networks of sensorimotor intentionality and appears across species. However, human infants begin to not merely respond to the direction of gaze of a caretaker but to initiate the joint attention (IJA) with the caregiver. Heather A. Henderson and Peter C. Mundy work primarily in developmental cognitive neuroscience rather than attachment research, but their analyses of the process by which infants initiate

---

[53] Jude Cassidy, "The Nature of the Child's Ties," in *Handbook*, p. 8.

[54] Heather A. Henderson and Peter C. Mundy, "The Integration of Self and Other in the Development of Self-Regulation: Typical and Atypical Processes," in Karen Caplovitz Barrett, Nathan A. Fox, George A. Morgan, Deborah J. Fidler, Lisa A. Daunhauer, eds. *Handbook of Self-Regulatory Processes in Development New Directions and International Perspectives: The Integration of Self and Other in the Development of Self-Regulation* (Published online by Routledge on: 17 Dec 2012), p. 117.

joint attention clearly touch upon the role of attachment and inner working models in facilitating exploration:

> This increasing capacity to *initiate* joint attention bids (IJA) with social partners allows for the shared processing and experience of objects and events in the environment. This more gradually developing ability builds on the more reflexive attention sharing skills characteristic of earlier infancy and reflects more intentional and flexible attention engagement/disengagement processes as well as social cognitive developments including representations of others' perspectives.[55]

Initiating joint attention is just one example of the role of internalized models for action constructed as interacting systems of ever higher-order cortical-subcortical networks that, based upon regularities observed in the world and the body, build upon themselves to articulate subcortically encoded primary dispositions. The infant's internal working model, which gradually incorporates not just strategies for response but also observed modes of action on the world, is the core of the self. As the infant gains mobility, its cortical systems mature in coordination with this expanded range of perception and action, and as its model of the world grows more complex, the emerging self grows correspondingly more complex.

The relationships among attachment, exploration, and wariness (which becomes an increasingly important factor as the amygdala becomes disinhibited from 6 to 9 months) seem to stabilize in the period between 12 and 18 months, which is considered a sensitive period. Indeed, the standard assessment of attachment quality—Mary Ainsworth's "Strange Situation" scenario—is applied to 12- to 18-month-old infants and effectively combines all three elements to observe the way in which attachment supports exploration and allays wariness. This classic test begins with the assessor bringing the caregiver (usually the mother) and the infant into the experiment room; (1) the assessor then leaves so that the caregiver and the infant are alone for an interval; (2) a stranger comes in;

---

[55] Henderson and Mundy, "The Integration of Self and Other in the Development of Self-Regulation," p. 119.

(3) the caregiver leaves; (4) the caregiver returns and the stranger leaves; (5) the caregiver leaves again, leaving the infant alone; (6) the stranger returns; (7) the caregiver returns and the stranger leaves. The assessor is most focused on the infant's reactions to the caregiver's return in intervals (4) and (7). Based on patterns of infant reaction, Ainsworth and her lab first developed three categories of attachment styles—secure, anxious-avoidant, anxious-ambivalent/resistant—but then added a fourth, disorganized/ disoriented. Needless to say, the normative style is "secure," in which the infant is initially distressed when the caregiver leaves but is happy and quickly reassured when the caregiver returns, is wary of the stranger when alone but friendly when the caregiver is present, and explores the toys in the room while checking in with the caregiver, who serves as a secure base.[56] What is remarkable is the robustness and predictive power demonstrated by this attachment assessment in longitudinal studies: the inner working models constructed by age 18-months remain core features of the self throughout life.[57]

While studying development after the age of 18 months largely becomes the domain of cognitive and social psychology, neuroscience certainly has much to contribute to our understanding of these processes. Neuroscience traces the maturation of neuronal systems like the default mode network and executive control networks and the progressive myelination and activation of the prefrontal

---

[56] Jeffry Simpson and Jay Belsky, among others, have pointed out that there may be an evolutionary logic to different features of the attachment styles that prepare young humans (and other mammals) for harsh environments:

> If maternal rejection was a valid proximal cue of the severity of future environments, avoidance tendencies might motivate children not only to move away from their parents earlier but also to become more opportunistic and risk′ taking, thereby facilitating survival and early reproduction in arduous environments.

Jeffry A. Simpson and Jay Belsky, "Attachment Theory within a Modern Evolutionary Framework," in Cassidy and Sharver, eds., *Handbook of Attachment*, p. 99. Also see the discussion of rats in H. Jonathan Polan and Myron A. Hofer, "Psychobiological Origins of Infant Attachment and Its Role in Development" in *Handbook of Attchment*.

[57] The best-known longitudinal study of the role of attachment styles is in the Minnesota *Longitudinal* Study of Risk and Adaptation. See L. Alan Sroufe, "Attachment and development: A prospective longitudinal study from birth to adulthood," *Attachment and Human Development* 7 (2005), pp. 34-80.

cortex as well as the cortical underpinnings of the development of episodic memory and language. The pieces are in place, as the fundamental logic of structuration is set by 18 months, even though much happens after 18 months: the language explosion, the development of theory of mind, empathy, and self-awareness, and the shaping of sexual identity. Much of this is strongly informed by environmental factors interacting with the neuronal structures readied for learning in early infancy, and this moves the inquiry to a new phase. Culture, social norms, the built environment all interact with the particular features of an infant's early development:

> Although the physiology of the brain and predictive processing are universal features of the species, the social environments and the networks of social practices in which people live are certainly not universal; they are highly variegated. Therefore, the higher-level modelling processes will not be universal either. Social environments are heterogeneous and social actors engage not only in differing practices, but the practices taking place in different lifeworlds are constrained by class, age, gender, ethnicity, sexual orientation, and geography. These different axes of social difference intersect with each other to create the rich and nuanced web of social life in any given society or culture.[58]

Where is the Self?

The difficulty in discussing the development of the neuronal self is that this "self" does not look like what one would expect. The self that begins to emerge in early infancy is a complex network of networks that ties increasingly complex action scenarios together with accumulating memories and valuations of a self and a world. These core networks—this self—are built upon what the experience-dependent cortical systems of this particular brain—with its individuated subcortically articulated dispositions—have learned about the particulars of this

---

[58] Michael P. Kelly, Natasha M. Kriznik, Ann Louise Kinmonth & Paul C. Fletcher, "The brain, self and society: a social-neuroscience model of predictive processing" *Social Neuroscience*, 14.3 (2019), p. 273.

body, these caregivers, and this lifeworld. In contrast, the *consciously accessible* self is just the highly mediated representation of some small portion of this larger, richer neuronal self. For this reason, I tend to view the concern for consciousness to be something of a distraction that unfortunately feeds into the old Cartesian understanding of the self as the *res cogitans*. Consciousness is decidedly a neat trick, and it also plays a vital role in a fourth tier of associational and attentional processing in the service of action, but it is such a high order phenomenon that it is a challenge to extract information about the structures and processes on which it operates from what we can observe through phenomenal self-reflection. My concern is not for this process of synthesis but on the structures that provide it with its material and shape its possibilities. The neuronal self, as structure, a system of systems, emerges slowly as the brain develops and is shaped both by the personal history behind this emergence and by the body within which it has grown.

At the beginning of the book, I proposed a process by which identity shatters only to return as it was, yet entirely different. The homunculus is gone; the *res cogitans* is gone. Any single coherent "thing" is gone. In their place, we have cross-communicating systems of memory, action, perception and assessment in the brain. Yet this neuronal system of systems is the experiential self that we know already, and it reproduces all the attributes as well as the problems of selfhood that we might hope to escape through scientific analysis. Our problems of inescapable self-contradiction and inadequate self-understanding, however, are real and part of the very structure of selfhood. Like the experiential self, the neuronal self offers no promise of the sort of full self-knowledge that we have failed to discover through millennia of trying. Nor will the neuroscientific model give us complete knowledge of our fellow humans. Nor does the neuronal self provide any easy ways out of the internal contradictions incorporated into the self from the contradictory nature of experience and of the primary intentionalities that shape the self as it emerges. Still, the neuronal model of selfhood at least gives us an account of what there is to know about the formation of the experiential self, even if there is no hope that we can fully know the crucial details. While the neuroscientific model of the self serves a vital *critical* function in providing constraints on a formal logic for the inner structuring of meaning, the task of a higher-order knowledge of the self falls not to neuroscience but to the human sciences—literature, history, psychology, anthropology, and sociology. Perception, cognition, emotion and memory all are intimately related within the cortical and subcortical dynamics that underlie them. Among these aspects of experience, there is neither absolute identity nor absolute difference: instead, they arise through the complex bottom-up and top-down, synthesizing participation in the neuronal networks we have considered. Similarly, as the proposed development of the neuronal self explored in this chapter suggests, the relation of individuals to their society cannot be reduced to absolute identity or difference: the choice between "nature or nurture" is not well formed.

An understanding of the neuroscientific accounts of perception, emotion, and development can help us not simply to unravel old conceptual dichotomies that don't accord with the realities of human experience but to replace them with mediating structures that offer a more complex, nuanced relationship between the elements of our experience. In the next chapter, I extend my account of the neuroscientific model of the complexity of interaction between primary intentionality, perception, and emotion to memory and the semantic structures of language because neuroscience provides a compelling critical framework for our discussions of memory and language in the humanities and especially in my own field of literary studies.

CHAPTER SIX

Building a World of Meaning: The Neural Dynamics of Memory

We do not—and cannot—live entirely in the present. We see the world and all the objects and events in it not simply as they "are" but through comparing the present moment with past experience. We have no choice: this is how we are built. The past inheres in and shapes our encounters with the world from the most basic levels of sensory perception to the highest forms of self-reflection. From an evolutionary perspective, using the patterns of the past to respond effectively to the present is precisely the point of the brain. To this end, the brain captures patterns extracted from the neuronal representations of past experience at every level. We have seen how the primary sensory cortices organize their neuronal networks according to the regularities in the data that our sight, hearing, touch, taste and smell provide. We also have seen how the salience networks that shape attentional focus within the sensory cortices draw on assessments of what in experience has proved painful, rewarding, or interesting. It is through the assessment of experience that the self emerges.

Memory in the brain, at its most basic, is the extraction of patterns from experience through synaptic reorganization. Although theoretical models for this reorganization have grown in sophistication since the introduction of the simple neural networks driven by Hebb's learning rule, the core idea has not changed: memory is embodied in the weighting matrix that represents the strength of the synaptic connections in the neural network. Learning is in the updating of that weighting matrix. I stress this basic principle because throughout this chapter, the challenge in discussing the nature of memory invariably returns to the questions of, first, how the synapses are selected to be strengthened and, then, how they are strengthened. While I have considered issues in the training of neural networks in Chapter 1, the most basic and general form of memory in the brain, in this chapter my focus will shift to the higher-order types of memory usually termed "episodic" and "semantic" memory.

Episodic memory is memory of personally experienced events, while semantic memory refers to the structures of meaning that one accumulates around "objects," a loose category that includes physical objects, actions, concepts, and anything else that can be abstracted as a component of experience. However, even the seemingly simplest event and the objects participating in it—seeing a red ball partly hidden by a living room chair—are internally complex and reflect the activation of many neuronal systems in the brain. As discussed in Chapter 2, the basic act of seeing entails attentional control of foveation; visual perception uses feedback activation from an "object store" (that identifies the object in the visual field as a *partially* seen *ball*. The attention system uses assessments of salience and value that draw on memories (both episodic and semantic) about balls in general and that ball and that place in particular. Valuation in turn draws on affective associations and action plans and requires the participation of high-level neuronal networks that retrieve and weigh these values and integrate them into current concerns. An "event" binds together all of these patterns of activation of neuronal systems that articulate the quality and import of the experience. The memory of this event therefore correspondingly preserves these bindings across systems of neuronal networks, and recalling this memory entails the selective activation of the networks used to preserve the coherent structure of the event.[1]

This neuroscientific account of the task of memory—of what a memory *is*—matches our general sense of memory: memory is bound to objects yet

---

[1] As Patricia Bauer reminds us, memory in the brain

is nothing like a file cabinet, and there are no file folders with records of experience. Rather, memory representations are made up of individual elements of experience that are encoded in synaptic connections between individual neurons in various regions of the cortex.... To live on as "memories," the patterns of activity must be stabilized and integrated, processes carried out by a multi-component neural network that includes structures in the neocortex and the medial-temporal lobe.

Patricia J. Bauer, "Development of episodic and autobiographical memory: The importance of remembering forgetting," *Developmental Review* 38 (2015), p. 152.

ramifies endlessly. The neuroscience contributes a formalized biological architecture that supports this experiential, phenomenal understanding of memory.[2] Yet the conceptual underpinnings nonetheless shift in this reframing. Terms like "objects," "events," and more crucially, "meaning" and "self" understood in relationship to the biology of memory take on powerful new dimensions. The goal of this chapter is to provide enough of an account of the neuroscientific model for the dynamics of memory to clarify the implications for our broader understanding of ourself and our construction of meaning in the world.

The Basic Model

The effort to resolve a problem in the fundamental behavior of neural networks that was discovered early in the resurgence of connectionist models has shaped thinking about memory for the past twenty-five years. This problem is the seemingly abstract one of *catastrophic interference*, in which an artificial neural network's learning of one series of facts was catastrophically impaired by its learning a second sequence. Michael McCloskey and Neal J. Cohen first introduced this phenomenon as proof that neural networks could not serve as an adequate model for cognitive functions.[3] However, James McClelland, in a famous experiment, demonstrated a solution to the problem. His team trained a supervised artificial neural network on a set of patterns describing living things using sentences like:

> Robin can grow, move, fly.
> Oak can grow.
> Salmon has scales, gills, skin.
> Robin has wings, feathers, skin.
> Oak has bark, branches, leaves, roots.

---

[2] There is a well-developed field of neurophenomenology that engages with questions derived from Edmund Husserl's phenomenology that centers in particular on conscious experience, which is outside the focus of this book. For a convenient discussion of issues surrounding neurophenomenology, see Liliana Albertazzi, "Naturalizing Phenomenology: A *Must Have?*" *Frontiers in Psychology*, published 22 October 2018, doi: 10.3389/fpsyg.2018.01933.

[3] Michael McCloskey and Neal J. Cohen, "Catastrophic Interference in Connectionist Networks: The Sequential Learning Problem," Psychology of Learning and Motivation 24 (1989), pp. 109-65.

On looking at how the network divided the object space, McClelland and his associates discovered that the network developed the following taxonomy for the terms:

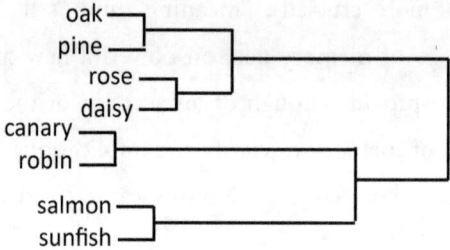

Figure 6.1 An Artificially Generated Dendrogram[4]

Rumelhardt's team then taught the system about penguins:

> Penguin can swim, move, grow.
> Penguin has wings, feathers, skin.

Penguins, needless to say, violated the rules of the world as the network knew them. When the previously stabilized system was trained over and over with just the penguin information, catastrophic interference arose to seriously disrupt the earlier taxonomic patterns that defined the identities of the components. However, McClelland's team also discovered that such interference could be greatly restricted if the new training set contained both the penguin information and the old data.[5] That is, if one gradually presented the new information into the system along with a rehearsal of the old patterns, the neural network incorporated the new rules into its representational schema with far less disruption of the old patterns, even if learning about the particular rules for penguins is less effective than in the approach of simply repeating just penguin facts.

McClelland's group argued that an already established model of comple-

---

[4] The length of the lines indicates the relative closeness of the categories. Thus, the rose/daisy inclusive group is closer to the pine/oak than the canary/robin is to the salmon/sunfish. Derived from James L. McClelland, Bruce L. McNaughton, and Randall C. O'Reilly, "Why There Are Complementary Learning Systems in the Hippocampus and Neocortex: Insights from the Successes and Failures of Connectionist Models of Learning and Memory," *Psychological Review* 102.3 (1995): 419-57.

[5] McClelland, et al., "Complementary Learning Systems," p. 435

mentary episodic and semantic memory systems of the brain serve precisely to allow potentially destabilizing new information to be incorporated into the structures of memory without catastrophic interference. They proposed a theoretical model for the relationship between the two forms of memory: the brain had mechanisms to quickly record the components of an event through *episodic memory* that were complemented by a much slower process that constructed *semantic memory* through extracting the regularities discovered within events.[6] This second phase of integrating the details of episodic memory into memory in a way that does not disrupt semantic structures is called *systems consolidation*. While Larry Squire and Pablo Alvarez, to account for observations about retrograde amnesia, had earlier proposed a model of systems consolidation as part of a complementary learning system that translated episodic memory into semantic memory,[7] McClelland's group, in demonstrating systems consolidation as the solution to catastrophic interference, identified a core function of these complementary learning processes.[8]

The basic design for the complementary memory systems is straightforward. Individuals have many experiences during the day; some are trivial and "unmemorable" while others have a greater impact. Under the Hebbian rule that "neurons that fire together wire together," the distinctive patterns of activations in the widely distributed neuronal systems that process these events in the brain leave traces in the form of slight changes in the strength of the participating synapses. The hippocampus and its associated cortical regions in the medial

---

[6] Endel Tulving in 1972 proposed the complementary memory model in "Episodic and semantic memory," in Endel Tulving and Wayne Donaldson, eds., *Organization of Memory* (New York: Academic Press, 1972). pp. 381–402.

[7] Larry Squire and Pablo Alvarez, "Retrograde amnesia and memory consolidation: a neurobiological perspective," *Current Opinion in Neurobiology* 5 (1995), pp. 169-77.

[8] This systems-consolidation framework has held up remarkably well throughout all the advances that have been made in the past two decades. Researchers have pointed out that the sort of statistical learning used to organize semantic memory networks need not depend only on rehearsal of memories first encoded in the episodic memory but also can use regularities directly encountered in experience even in the absence of a well-functioning episodic memory system (particularly in early infancy). This addition is important but does not undermine the basic systems consolidation model.

temporal lobe somehow record these residual patterns, mark the important sets of patterns for preservation, and tag the rest of the synaptic changes for deletion. Sleep then plays a role in this preservation and deletion and also begins the process of using the regularities in events to update semantic memory, the brain's model for how the world and the self work.[9] Within this general framework, there are huge unknowns about core processes. Researchers reporting on their work often begin with disclaimers about the current state of understanding:

> A fundamental cognitive process is to map value and identity onto the objects we learn about. However, what space best embeds this mapping is not completely understood.[10]

> Although a subject of intense study, the fine-grained mechanisms underlying how we retrieve episodes of experience are unknown.[11]

> An accumulating body of evidence has demonstrated that sleep is beneficial for various types of learning and memory. However, the underlying neural mechanisms for how sleep impacts on learning and memory are poorly understood.[12]

> The hippocampus is a brain region supporting memory, but the network-level operations that continuously incorporate new experiences, segregating them as discrete traces while enabling their interaction, are unknown.[13]

---

[9] See, for example, Wei Li, Lei Ma, Guang Yang and Wen-Biao Gan, "REM sleep selectively prunes and maintains new synapses in development and learning," *Nature Neuroscience* 20.3 (March 2017):427-37.

[10] Evelyn Tang et al., "Effective learning is accompanied by high-dimensional and efficient representations of neural activity," *Nature Neuroscience* 22 (July 2019), p. 1000

[11] G. Elliott Wimmer et al., "Episodic memory retrieval success is associated with rapid replay of episode content," *Nature Neuroscience* 23.8 (August 2020), p. 1025.

[12] Masako Tamaki et al., "Complementary contributions of non-REM and REM sleep to visual learning," *Nature Neuroscience* 23.9 (September 2020), p. 1150.

[13] Giuseppe P. Gava et al., "Integrating new memories into the hippocampal network activity space," *Nature Neuroscience* 24.3 (March 2021), p. 326.

Still, the speed and synergistic character of developments in the neuroscientific understanding of all the components of memory encoding and retrieval has been remarkable. In particular, researchers have made great strides in elucidating the broader systemic interactions that support memory. This broader logic of construction involved in both encoding and retrieval is of central importance for our thinking about experience and meaning. In the pages that follow, I introduce the main brain structures and operations that participate in the processes of memory. The details are a bit complex, but they are important for understanding the evolving conceptual framework for the construction of long-term memory from its many disparate parts.

Such Stuff as Dreams are Made on

Experience is not simple. In visual experience, we begin with the spiking neurons in V1, already influenced by the attentional systems that draw on the affective weight of recalled memories; we have the objects and their immediate associations—once again made available through recall processes—and we have all the simultaneously available sensory data from other modalities, synthesized in multimodal association cortices. We have the internal contexts within which the event is assessed, including current plans, concerns, and physical and affective states. We have the external environmental contexts: in many species there are specialized brain regions that have evolved to track spatial cues, and in humans these regions have perhaps been generalized to track "place" more broadly conceived. Moreover, the brain has to record temporal succession, how experiences unfold over time.

The brain records the basic material of experience throughout the day simply in the process of neuronally processing the flow of experience. That is, synaptic activations approximately follow the Hebbian learning rule that "neurons that fire together wire together," so that the processing itself changes synaptic weights. Yet, in the flux of experience, we see many objects and events and ways of connecting objects to one another and to events (the blue pen is on the table; I dropped a red one on the floor by the leg of the desk, and on and on). Since the goal of the memory system that subserves predictive processing is to build a model of the world and the self that allows one usefully to predict states

as the probable cause of the current input, the learning processes must separate incidental events from those that reflect persistent patterns[14] To train the system usefully, the data should not be an unfiltered stream of neuronal activity but should be biased toward *significant* states, the sorts of states tagged as episodic memory. Since the "stuff" of an episodic memory are patterns of neuronal network activations passed layer by layer, strengthening the synaptic connections between layers (say, between V4 and V1) in the encoding of an episodic memory importantly trains the system on "weighted" statistical significance and makes useful predictive processing possible.[15] Thus, it turns out the manner in which memory encoding and recall involve the primary sensory cortices significantly links the weighting processes for memory consolidation to the seemingly quite separate framework of cortical model building required for predictive processing paradigms.

From the neuroscientific perspective, experience stays with us in complex ways. Perception, feeling, and planning are all intertwined and mutually shaping in the constant flow of lived experience, and when moments in time particularly matter, the neuronal patterns of activation that subserve those experiences are inscribed in episodic memory but also in the very texture of how we engage the world.

Memory and the Medial Temporal Lobe

One extremely important form of long-term memory lies in the low-level synaptic dynamics needed for the pervasive predictive processing by the neuronal networks in the neocortex. The training of neural networks to participate in predictive processing proceeds layer-by-layer: for example, even in the early visual

---

[14] See Chapter 1.

[15] While I focus on episodic memory, in fact even before the circuits in the hippocampus thought to support episodic memory become mature, the developing infant has mechanisms for tagging complex syntheses of neuronal input via prefrontal lobe processing to achieve this effect of training through weighted statistical learning. See Gal Raz and Rebecca Saxe, "Learning in Infancy Is Active, Endogenously Motivated, and Depends on the Prefrontal Cortices," *Annual Review of Developmental Psychology* 2020.2, pp. 247-68. I discuss this article in Chapter 5.

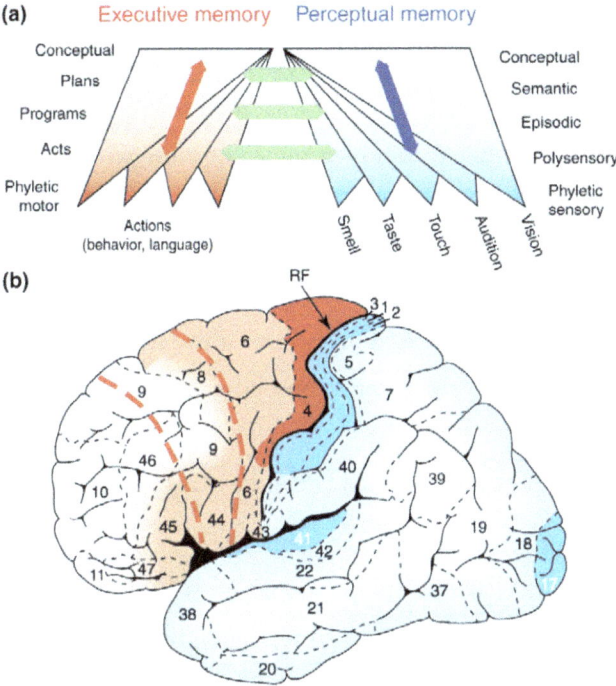

Figure 6.2: Distribution of Memory in the Brain[16]

Areas shaded in red play a role in memory for planning action: the deepest red area (Brodmann area 4) is the primary motor cortex. Similarly, blue areas preserve memory associated with perception: the deepest blue are the primary somatosensory (BA 1,2, and 3), visual (BA 17) and auditory (BA 41) cortices. The lighter the color, the more abstracted, synthesized, and higher order are the memory networks. RF = the Rolandic fissure (central sulcus).

system, V4's feedback activation to V1 must become trained such that if a pattern is the input of V1, V4 can predict (in terms of its representation space) what is the most likely cause for the input from V1, which it then feeds back to V1. The inscription of memory at the synaptic level—layer by layer—encodes a highly dispersed model of the world and the self. This training of neuronal connectivity based on statistical regularities and assessed significance is a type of what

---

[16] Joaquin Fuster, "Upper Processing Stages of the Perception-Action Cycle," *Trends in Cognitive Science* 8.4 (April 2004), p. 144.

cognitive scientists call *implicit* long-term memory, long-term memory not available to cognitive access.[17] Its counterpart is *explicit memory*, memory accessible to conscious recall, which is what we usually think of "memory" in ordinary conversation. Explicit long-term memory, also referred to as *declarative* memory, usually further divides into episodic and semantic memory.[18]

While neuroscientific approaches are perhaps outgrowing the standard taxonomy of memory that was based on cognitive science and neurology, "episodic memory" remains a useful concept with which to begin an account of the general framework for the systems consolidation model of memory.[19]

The Medial Temporal Lobe Memory System

The "hardware" for recording episodic memories is in the medial temporal lobe (MTL), which includes the parahippocampal gyrus (perirhinal cortex, the parahippocampal cortex, the entorhinal cortex) and the hippocampus itself. The hippocampus is part of the so-called allocortex and has structural features that

---

[17] In this chapter I focus on two forms (implicit and explicit) of *long-term* memory, relatively stable changes in synaptic connectivity that persist over time. In contrast, there are also (1) very short-term "sensory memory" that buffers low-level sensory data as a stream, (2) "short term memory" that "chunks" data from the input stream into slightly larger units for processing, and (3) "working memory" that serves as a buffer of higher-order data for use in a wide range of cognitive tasks. These three types of memory are part of the mechanisms for cognitive processing rather than elements that define the representational structure of the world of data that is processed.

[18] Although I rely on this terminology here, advances in the neuroscience of memory are leading to calls for rethinking the common taxonomy for memory. For example, Sinéad L. Mullally and Eleanor A. Maguire argue:

> While inferring hippocampal functionality in the infant through cognitive testing and inference alone may have been appropriate three decades ago (…), surely in this era of advanced, non-invasive neuroimaging techniques we should be attempting to ask more sophisticated and potentially more useful questions of this developing system. In this way, the infant memory literature could be unshackled from terms like declarative, explicit and conscious memory, terms which have caused theoretical divides within the literature.

"Learning to remember: The early ontogeny of episodic memory," *Developmental Cognitive Neuroscience* 9 (2014) pp. 23-24.

[19] Other routes into semantic memory that do not rely on episodic memory prove to be extensions and elaborations of this framework rather than contradictory findings.

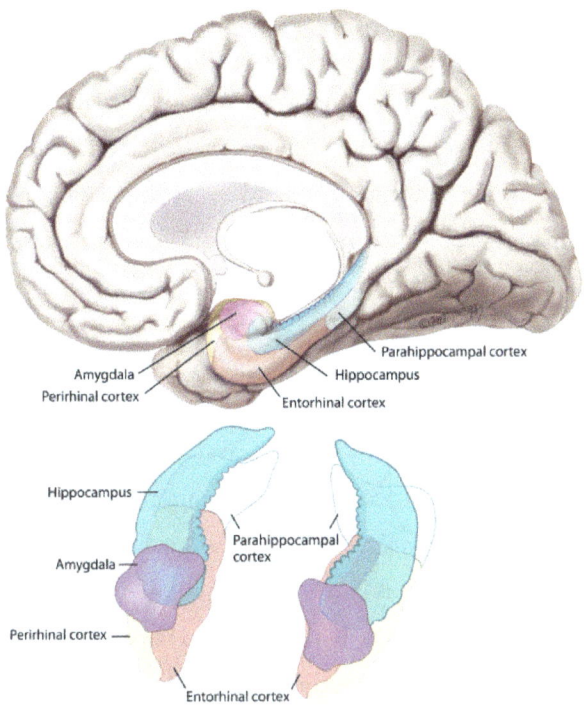

Figure 6.3: The Medial Temporal Lobe Structure – Parahippocampal Cortex, Entorhinal Cortex, and Hippocampus[20]

distinguish it from the neocortical regions to which it is connected. For example, it has only three main layers of neurons rather than the neocortical six. Specific aspects of these structural differences, to be discussed below, make possible the specialized functional features that support the encoding and retrieval of long-term episodic and semantic memories.

The organization of the medial temporal lobe is significant for its unique, evolved specializations that allow the brain to bind the details of events with their contexts and their temporal sequencing into episodic memory traces (engrams). The processing in the MTL is complex, which means that a "memory" is more

---

[20] F.D. Raslau, I.T. Mark, A.P. Klein, J.L. Ulmer, V. Mathews and L.P. Mark, "Memory Part 2: The Role of the Medial Temporal Lobe" *American Journal of Neuroradiology* May 2015, 36 (5) 846-849; DOI: https://doi.org/10.3174/ajnr.A4169, Figure 1, Adapted with permission from Purves D, Brannon E, Cabeza R, et al. *Principles of Cognitive Neuroscience* (Sunderland, MA: Sinauer Associates, 2008).

than a stitching together of synaptic activations spread across the brain. Instead, an episodic memory is a synthesis that transforms the data to create component activations to be stored. Some components are new, while some match pre-existing historical data, so that the array of patterns of activations stored as a memory is in part unique to the moment and in part framed by earlier experience.

The different areas in the medial temporal lobe have distinct functions. While these functions are a matter of intense research, I present here a simplified consensus account.

The Parahippocampal Gyrus

In looking at the multi-step process of synthesizing episodic memories, we begin with the perirhinal, parahippocampal, and entorhinal cortices that comprise the parahippocampal gyrus, which is part of the temporal lobe. The perirhinal cortex synthesizes information about the particular objects and other types of sensory data that participate in an event. It receives input from the insular, orbitofrontal, and anterior cingulate cortices, along with the temporal association cortices and the amygdala. Thus, the perirhinal cortex receives information not just about the objects and other elements of an event (from the association cortices) but also about their salience and affective valences (from the frontal and insular cortices and the amygdala). The perirhinal cortex projects its activations both to the lateral entorhinal cortex and to the parahippocampal cortex.[21]

In contrast to the perirhinal cortex, the parahippocampal cortex synthesizes "place" and context information coming from the retrosplenial and posterior parietal cortices with additional "object and pattern information" from the perirhinal cortex. In experimental animals like rats, the parahippocampal processing is quite literally about specific places, but in humans, this synthesis appears to broaden to include other types of invariant aspects that frame events.

---

[21] See the discussions of the relations of the perirhinal and parahippocampal cortices to the entorhinal cortex in Eirik S. Nilssen, Thanh P. Doan, Maximiliano J. Nigro, Shinya Ohara and Menno P. Witter, "Neurons and networks in the entorhinal cortex: A reappraisal of the lateral and medial entorhinal subdivisions mediating parallel cortical pathways," *Hippocampus*. 29 (2019):1238–1254,"

The parahippocampal cortex reciprocally projects to the perirhinal cortex as well as to both the lateral and medial entorhinal cortex.

The entorhinal cortex provides the majority of the input to the hippocampus and serves to begin the synthesis of input from the parahippocampal cortex and a wide range of other cortical networks. The *medial* entorhinal cortex (MEC), for example, receives input from the retrosplenial cortex (a secondary association cortex for spatial data). The lateral entorhinal cortex (LEC) is strongly connected to the perirhinal cortex and combines its activations with input from the orbitofrontal, medial prefrontal and insular cortices. Given this complex connectivity and fMRI data on patterns of activation, Nilssen et al. propose that the LEC "provides the hippocampus with a highly integrated, multidimensional representation of sensory information, including changes over time, constituting the content of an episodic memory."[22]

The Hippocampus

The entorhinal cortex projects to the hippocampus through two separate pathways. First is the *trisynaptic* (indirect) pathway in which the entorhinal cortex sends patterns of activation to the dentate gyrus, which then sends its significantly processed activations to CA3 ("CA" stands for the Horn of Ammon [*cornu ammonis*]), which then sends its yet further processed activations to CA1. The second pathway is a *monosynaptic* (direct) pathway that directly connects the entorhinal cortex to CA1. It is in the trisynaptic pathway that the truly hard work of synthesizing individual episodic memories is accomplished.

The Dentate Gyrus

The dentate gyrus, the first stop in the trisynaptic pathway, has a set of specialized structures that produce its distinctive functional properties. The most remarkable are the *granule cells* (GC), small neurons that continue to be generated from stem cells throughout one's life. Because their resting membrane potential is hyperpolarized, it takes significant activation at their dendrites to cause spiking.

---

[22] Nilssen et al., "Neurons and networks in the entorhinal cortex," p. 1247.

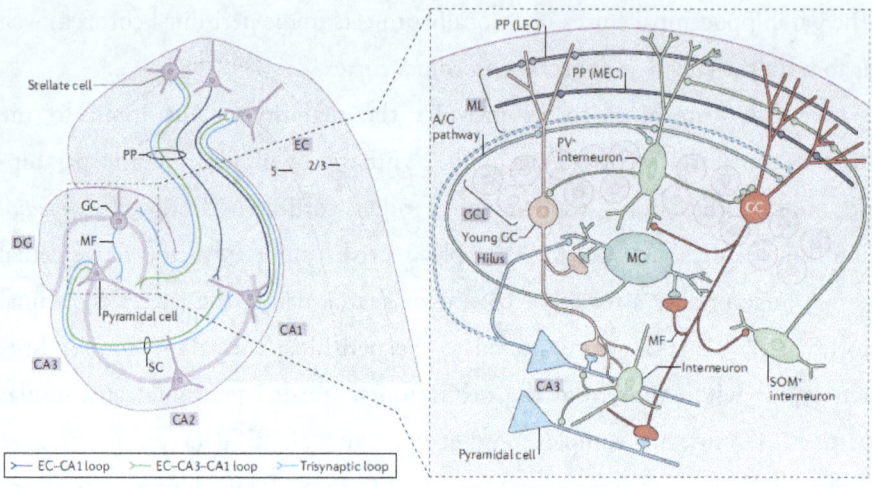

Figure 6.4: The Two Pathways in the Hippocampus

DG = Dentate Gyrus, MC = mossy cells, MF = mossy fiber axons, PP = perforant path input from EC, SC = Schaffer-collateral projections to CA1, A/C = associational and commissural pathway in CA3 [23]

However, there are about five times as many GCs as there are entorhinal cells, and the entorhinal neurons connect sparsely to the GCs (i.e., each neuron in the perforant pathway that connects entorhinal neurons to GCs connects to only a few GCs.) Additionally, there are a high number of inhibitory interneurons in the dentate gyrus.[24] The net result is a neuronal network of granule

---

[23] Thomas Hainmueller and Marlene Bartos, "Dentate gyrus circuits for encoding, retrieval and discrimination of episodic memories," *Nature Reviews Neuroscience* 21.3 (March 2020), p. 154. Associational collaterals in CA3 connect to neurons within the same CA3 region while commissural collaterals connect to the contralateral CA3.

[24] Beyond the projections from the entorhinal cortex, the dentate gyrus also receives modulatory and control inputs as well. The GCs of the dentate gyrus receive input from the supramammillary nucleus of the hypothalamus ("a key structure in the integration of cognitive and emotional aspects of goal-directed spatial learning behavior"), while the inhibitory interneurons and the so-called mossy cells of the *hilus* (the bottom layer of the dentate gyrus) receive neuromodulatory inputs of norepinephrine, serotonin, dopamine and acetylcholine along with feedback from CA3 and input from the granule cells. See Sebnum Nur Tuncdemir, Clay Orion Lacefield, and Rene Hen, "Contributions of adult neurogenesis to dentate gyrus network activity and computations," *Behavioral Brain Research* 374 (2019) 112112, p. 2. This article offers an excellent overview of the dentate gyrus.

cells with sparse connectivity and very low spiking rates.[25] However, despite this sparse activation, the granule cells have axons called *mossy fibers* with large specialized boutons that project to the pyramidal cells in CA3. Each mossy fiber connects to 10-20 CA3 pyramidal cells, but in each bouton there are many synaptic junctions so that when a granule cell does spike in what has been called a "detonator" synapse, it can induce the CA3 neurons connected to it to spike as well.[26]

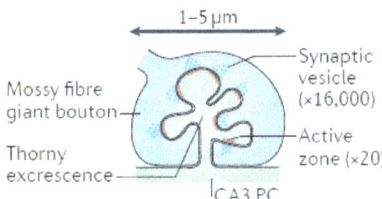

Figure 6.5: The Mossy Fiber- CA3 Synapse[27]

The functional result of this sparsely connected arrangement of low-firing neurons is a neural network capable of very strong *pattern separation*. That is, the patterns of activations from the entorhinal cortex that represent individual events will be held in separated, sparse patterns of activation within the dentate gyrus, a separation necessary to encode events in episodic memory as distinct patterns. In early neural network modeling and early studies of dentate gyrus activity, researchers believed that the patterns for each event as represented in the dentate gyrus were completely separate, but more recent evidence suggests that this separation is more complexly graded, with some overlap among shared elements

---

[25] By one measure, only 2% of the granule cells fire in response to a visual experience, in contrast to 18% of the pyramidal neurons in CA3 and 35% in CA1. Tuncdemir et al., "Contributions of adult neurogenesis to dentate gyrus network activity and computations," p. 3.

[26] Caroline Le Duigou, Jean Simonnet, Maria T. Teleńczuk, Desdemona Fricker and Richard Miles, "Recurrent synapses and circuits in the CA3 region of the hippocampus: an associative network," *Frontiers in Cellular Neuroscience* 7 (January 2014), article 282, p. 5.

[27] From Figure 2, Nelson Rabola, Mario Carta, and Christophe Mulle, "Operation and plasticity of hippocampal CA3 circuits: implications for memory encoding," *Nature Reviews Neuroscience* 18.4 (April 2017), p. 211.

in experiences.[28] As mentioned above, these differentiated patterns are then sent along the mossy fibers to CA3.

CA3

The CA3 subregion of the hippocampus receives projections from the dentate gyrus and the entorhinal cortex. Most crucially, however, its pyramidal cells have strong *recurrent collaterals* (branching axons) that connect with several thousand other CA3 neurons (both excitatory pyramidal cells and inhibitory interneurons).[29] The very distinctive recurrent organization of CA3 was identified in the original work of Cajal and has played a central role in thinking about the functioning of the hippocampus from the early days of connectionist neural network modeling. Recurrent connectivity introduced into neural network models creates *attractor networks* in which input patterns are strongly driven by the recurrent looping toward already stored patterns of connectivity that form attractor basins.[30] This dynamic leads to *pattern completion*, the aggregation of an input pattern to a stored, more complete, and stable pattern. Psychologists long before the era of neural network modeling had studied the human capacity for pattern completion in many forms and also knew of the importance of the hippocampus in forming episodic memories. So, connectionist modeling that identified the role of the recurrent network in CA3 provided an important clue in understanding *how* memory worked. However, neural network modeling also stresses that there always is the danger that important distinctive features of an event will disappear in the attractor network and thus there needs to be a way to

---

[28] See the discussions in Hainmueller and Bartos, "Dentate gyrus circuits for encoding, retrieval and discrimination of episodic memories" and Tuncdemir et al., "Contributions of adult neurogenesis to dentate gyrus network activity and computations." An additional useful account is offered by Alexa Tompary and Lila Davachi, "Consolidation Promotes the Emergence of Representational Overlap in the Hippocampus and Medial Prefrontal Cortex," *Neuron* 96 (September 27, 2017), pp. 228-41.

[29] Caroline Le Duigou, Jean Simonnet, MariaT.Teleńczuk, Desdemona Fricker and Richard Miles, "Recurrent synapses and circuits in the CA3 region of the hippocampus, p. 5. By one estimate, 30-70% of all CA3 pyramidal cell connections are to other CA3 cells (p. 1).

[30] See, for example, the discussion in Rabola et al., "Operation and plasticity of hippocampal CA3 circuits."

intervene in the attractor dynamics. The strong—if somewhat sparse—input from the dentate gyrus, which largely enhances pattern separation, serves precisely this purpose. CA3 receives the original input from the entorhinal cortex as well as the dentate gyrus' pattern-separated transformation of that input. "The winner of the competition between the information sent from the dentate gyrus and the stored representation from the CA3" enhanced by the direct entorhinal input is then sent to the CA1.[31]

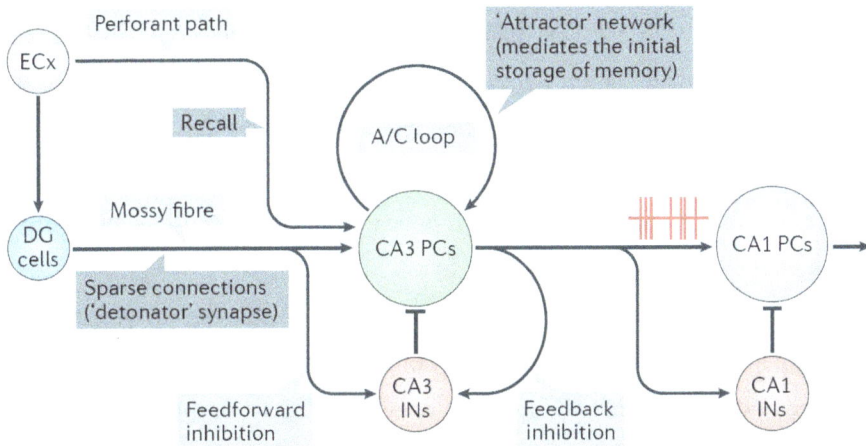

Figure 6.6: The Connections to CA3 in the Hippocampus[32]

The interaction between the dentate gyrus and CA3 in the trisynaptic pathways achieves remarkable results, even if there remain many details that are not yet fully understood. While the entorhinal cortex sends information about the objects, temporal sequence, and context of events, the processing in the trisynaptic pathway, with its tension between specificity (pattern separation) and generality (pattern completion) in encoding long-term memory, creates a flexible binding of objects, events, and contexts such that we can recognize similarities between events when specific elements change. This is no small accomplishment, when contrasted with the limitations of memories formed through the

---

[31] Shauna M. Stark and Craig E. L. Stark, "Introduction to Memory," Chapter 67 in Greg Hickok, Steve Small, eds., *Neurobiology of Language* (Academic Press, 2015), pp. 843.

[32] Rabola et al., "Operation and plasticity of hippocampal CA3 circuits: implications for memory encoding," p. 210.

monosynaptic pathways from the entorhinal cortex directly to CA1 or through cortical statistical learning that are discussed below.

CA1 and Subiculum

CA1 receives input from CA3 as the last stage in the trisynaptic pathway as well as direct projections from the entorhinal cortex in the monosynaptic pathway. In addition, it receives input from the *nucleus reuniens* region of the thalamus, which has strong connections to the medial prefrontal cortex. Output from CA1 has many pathways. On the one hand, CA1 projects to the entorhinal cortex, amygdala, prefrontal cortex, nucleus accumbens and other regions, but on the other, CA1 is strongly connected to the subiculum, which then also is connected to many regions in the cortex and subcortical nuclei.[33] Some researchers consider CA1 as the major output site for the hippocampus; others consider that output site to be the subiculum.

The function of CA1, at the intersection of the trisynaptic and monosynaptic pathways, appears to be complex. One proposal is that it serves yet one more comparative function, comparing CA3 output with existing memory engrams and then transmitting the results to the subiculum and cortical regions.[34] Reactivation of CA1, as the last stage in the trisynaptic pathway, also appears to play a prominent role in *memory consolidation* (as opposed to *memory formation*) during sleep, discussed below.[35] In particular, its connection to the medial prefrontal cortex seems crucial to the consolidation of episodic memories.[36] Because of the trajectory of hippocampal development, researchers trying to understand the causes of infantile amnesia have focused on the role of

---

[33] Arjun V. Masukar "Towards a Circuit-Level Understanding of Hippocampal CA1 Dysfunction in Alzheimer's Disease Across Anatomical Axes" *Journal of Alzheimer's Disease and Parkinsonism* 8.1 (2018), article 1000412.

[34] Stark and Stark, "Introduction to Memory," p. 843.

[35] Antoine R. Adamantidis, Carolina Gutierrez Herrera and Thomas C. Gent, "Oscillating circuitries in the sleeping brain," *Nature Reviews Neuroscience* 20 (December 2019):746-62

[36] Gareth R I Barker, Paul J Banks, Hannah Scott, G Scott Ralph, Kyriacos A Mitrophanous, Liang-Fong Wong, Zafar I Bashir, James B Uney & E Clea Warburton "Separate elements of episodic memory subserved by distinct hippocampal–prefrontal connections," *Nature Neuroscience* 20.2 (February 2017), pp. 242-50.

the monosynaptic pathway, which matures earlier than the trisynaptic pathway, in encoding memory. One suggestion is that the entorhinal cortex–CA1 loop supports statistical learning in early infancy in which relations are encoded as a single totality in an inflexible manner rather than as the combination of objects and contexts made possible by the trisynaptic loop.[37] In this model, infants can capture statistical regularities in semantic memory but not the details of the episodes from which the regularities are derived, thus resulting in the observed infantile amnesia.

The subiculum is between the entorhinal cortex and CA1. As noted above, CA1 primarily projects to the subiculum, which, in turn, is reciprocally connected to many regions in the cortex as well to subcortical nuclei. Much about this connectivity is poorly understood.[38] However, the subiculum has an important secondary function: it has a strong reciprocal connection to the hypothalamus. Through the subiculum, the *HPA stress system* has a strong impact on hippocampal function, and at the same time, the subiculum also can inhibit stress responses.[39]

In sum, patterns of activation go into the MTL memory system, which processes them, weaves them together, and creates a memory engram, a pattern of activations centered in the hippocampus but reciprocally projecting to all the cortices that contributed to the memory. In this model, memory is not so much

---

[37] See Cameron T. Ellis, Lena J. Skalaban, Tristan S. Yates, Vikranth R. Bejjanki, Natalia I. Córdova, Nicholas B. Turk-Browne "Evidence of hippocampal learning in human infants," *Current Biology* 31 (Aug. 2021), pp. 1-7; Sinead L. Mullally, "Commentary: Elucidating the neural correlates of early childhood memory," *International Journal of Behavioral Development* 39.4 (2015), pp. 306-07; and Anna C. Schapiro, Nicholas B. Turk-Browne, Matthew M. Botvinick and Kenneth A. Norman, "Complementary learning systems within the hippocampus: a neural network modelling approach to reconciling episodic memory with statistical learning," *Philosophical Transactions B of the Royal Society* (2016)) http://dx.doi.org/10.1098/rstb.2016.0049)

[38] See, for example, Yanjun Sun, Suoqin Jin, Xiaoxiao Lin, Lujia Chen, Xin Qiao, Li Jiang, Pengcheng Zhou, Kevin G. Johnston, Peyman Golshani, Qing Nie, Todd C. Holmes, Douglas A. Nitz and Xiangmin Xu, "CA1-projecting subiculum neurons facilitate object–place learning," *Nature Neuroscience* 22.11 (November 2019), p. 1857.

[39] Shane O'Mara, "The subiculum: what it does, what it might do, and what neuroanatomy has yet to tell us," *Journal of Anatomy* 207 (2005), p. 279.

constructed as drawn together and synthesized. However, the hippocampus is constantly at work, and its constant stream of syntheses are ephemeral unless those events that are marked as significant are then preserved through a strengthening of those engrams. This is the work of memory consolidation, which mostly happens during sleep.

Sleep and Memory Consolidation

The Complementary Learning Systems model of memory encoding functions in two steps. First, the hippocampus rapidly encodes a pattern of activations across widely distributed neuronal networks to represent an event. Then, replay of the hippocampus-centered patterns through system consolidation slowly trains the neocortical networks to add the new aspects of the event to their larger neuronal systems of representations that model the world and the self. As mentioned above, the development of the CLS model had two motivations. First was to account for the nature of retrograde amnesia after hippocampal damage: the famous patient HM could not remember recent events but could recall people, places, and facts, as well as episodes from his early life. This was evidence that episodic memory and semantic memory were distinct systems, even though semantic memory clearly was at least partly based on information captured through episodic memory. The second motivation was then to account for the relation of episodic to semantic memory: new relations that were revealed though episodic memory potentially could conflict significantly with the current structure of relationship articulated in the sematic memory system and could cause *catastrophic interference* within that structure. However, neural network models showed that the gradual incorporation of the new patterns into semantic memory through slow but repeated reactivations avoided this destabilization.

Scientists and even popular culture have long been aware that sleep helps with memory. After the discovery of REM (rapid eye movement) sleep, careful delineation of the sleep cycle helped researchers differentiate sleep's contributions to the retention of learning. Studying regional patterns of brain activity during sleep revealed strong activity by the hippocampus, which neatly fit the Complementary Learning System model and provided a mechanism for systems consoledation. Years of research exploring the patterns of brain activity during sleep have filled in many details and given a more fine-grained picture of how the consolidation of memory during sleep might work, although much also remains a mystery. The story is complex, so the first step is to introduce the basics of the sleep cycle.

THE SLEEP CYCLE

Any account of the sleep cycle must begin with tonic brain oscillation frequencies that characterize the different levels of sleep.[40] When we are awake, a beta frequency (12-30 Hz) predominates. This is a frequency fast enough to accommodate all manner of communications between neuronal networks throughout the brain. Given the heterogeneity of the signaling, the amplitude of the beta signal is low. When we begin to get drowsy, we shift into an alpha frequency (8.5-12 Hz): we are awake but not attending so carefully to our surroundings. When we slip into sleep, we first go through four stages of *NREM* (non-Rapid Eye Movement) sleep (although Stages 3 and 4 usually are considered together as Slow Wave Sleep) in which the frequency gets progressively slower while the amplitude gets larger. During NREM sleep, the levels of various neuromodulators—dopamine, noradrenaline, and serotonin—also drop. Stage 1 NREM sleep is a mix of waves of alpha frequency and of the slower theta frequency (4-7 Hz). The thalamus has not yet shut down most transmissions of activations from the senses (recall that vision first goes to the lateral geniculate nucleus of the thalamus while hearing is processed in the medial geniculate nucleus, and so on.) Some speech, like one's name, will register. Stage 1 NREM sleep, however, occupies just 5% of our sleep. We then shift into Stage 2, which accounts for 45-55% of sleep. During Stage 2, k-complexes (bursts of large-amplitude oscillations) and sleep spindles (short bursts of 12-15 Hz waves)

---

[40] There are many resources for exploring the neurobiology of sleep. For my account, I have drawn in particular upon the overview in Andrew S. Tubbs, Hannah K. Dolish, Fabian Fernandez, and Michael A. Grandner, "The basics of sleep physiology and behavior" in Michael Grandner, ed., *Sleep and Health* (Cambridge: Academic Press, 2019). I also have relied on the more focused discussion in Jens G. Klinzing, Niels Niethard and Jan Born, "Mechanisms of systems memory consolidation during sleep," *Nature Neuroscience* 22 (October 2019), pp. 1598-1610.

appear. As discussed below, these sleep spindles are believed to play a role in memory consolidation. During the first cycles during the nightly sleep cycle, Stage 2 usually transitions to Stage 3 NREM sleep, which, together with Stage 4, is also called slow wave sleep (SWS) since the brain drops to a delta wave (about 1 Hz) oscillation, also called a slow oscillation (SO). Sleep spindles are yet more frequent in SWS, and one additional component of these spindles are short periods of high-frequency (80-140 Hz) ripples. During Stages 3 and 4, the thalamus does not transmit sensory data to the sensory cortices, although it does send other activations. Stages 3 and 4 then transition to REM sleep.

During REM sleep, acetylcholine levels rise, which results in a pattern of high-frequency, low amplitude oscillations that comes close to the state when we are awake. At the same time that most muscle activity is blocked by brainstem nuclei, the muscles controlling the eyes make them dart about, giving this sleep stage its name. We dream in both REM and NREM sleep, but because of the heightened activity across the brain during REM sleep—including in the primary sensory cortices—REM dreams are much more vivid when recalled. REM sleep shifts back into NREM Stage 1 sleep, and this cycle is repeated from four to six times during the night, with REM sleep totaling about 25% of sleep.

What drives the sleep cycle? Because sleep is so important and sleep problems so widespread, much research has gone into clarifying the biology. While answers are still slow in coming because new levels of interaction and complexity are being discovered,[41] the current standard account is the so-called Two Process Model of sleep, developed in the early 1980s.[42] The first process, "Process C," is driven by the circadian pacemaker, where the primary control point is the suprachiasmatic nuclei (SCN) of the hypothalamus, which receives input from the retinas. The second component is "Process S," a homeostatic feedback loop for controlling sleep-wake cycles. While all the markers that drive

---

[41] Clifford B. Saper and Patrick M. Fuller, "Wake-sleep circuitry: an overview," *Current Opinion in Neurobiology* 44 (2017), pp. 186-192, offers a good example of research that continues to complicate the current models.

[42] For a review of this model, see Alexander A. Borbély, Serge Daan, Anna Wirz-Justice and Tom Deboer, "The two-process model of sleep regulation: a reappraisal," *Journal of Sleep Research* 25 (2016), pp. 131-43.

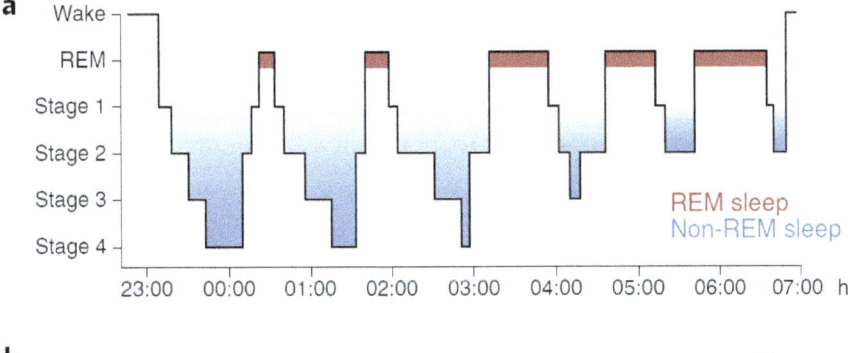

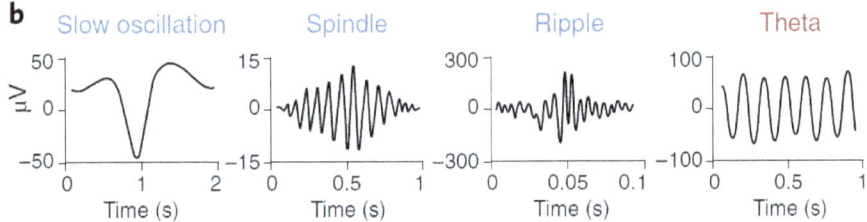

Figure 6.7: The Sleep Cycle[43]

the homeostatic regulation of sleep are not fully understood, most researchers point to the accumulation of adenosine in the extracellular environment of the brain: as adenosine, a byproduct of the cellular utilization of adenosine triphosphate (ATP) as a source of energy, accumulates, the brain is pushed toward sleep. During sleep, and especially during slow-wave sleep, it is thought that the adenosine and other metabolic by products are cleared and ATP replenished. Exactly how Process C and Process S interact remains a subject of intense research, but among other processes, one suggestion is that the accumulation of adenosine reduces the release of the neurotransmitter glutamate in the SCN's response to light information from the retina.[44] There are then two pathways involving other regions in the hypothalamus, with the pons and other nuclei in the brainstem. These neuronal networks project to the thalamus to activate neurons that generate wakefulness. At the same time, another nucleus in the hypothalamus,

---

[43] Klinzing, Niethard and Born "Mechanisms of systems memory consolidation during sleep", p. 1603.

[44] Tom Deboer, "Sleep homeostasis and the circadian clock: Do the circadian pacemaker and the sleep homeostat influence each other's functioning?" *Neurobiology of Sleep and Circadian Rhythms* 5 (2018) pp. 68–77.

the ventrolateral preoptic (VLPO) area, actively produces inhibitory neurotransmitters that promote sleep and decrease the activity in those areas that generate wakefulness. While the details are complex and not yet clarified, the central point for my purposes is that once again we encounter a distributed set of networks operating through multiple channels and driven by many inputs that interact to produce what we know as sleep and wakefulness.

Sleep is an evolutionarily old biological process. Some birds, for example, undergo REM and NREM sleep, with slow-wave activity in NREM sleep that is similar to mammalian behavior.[45] Yet the loss of vigilance that comes with sleep is expensive, even though some mammals and birds sleep with only one hemisphere at a time to minimize the cost.[46] Why then sleep at all? As Tononi and Cirelli stress, given the cost of sleep, it must be important. Their explanation is that "sleep is the price of plasticity." That is, the process of acquiring long-term memories requires it.

Memory Consolidation

While biological housekeeping is one key function of sleep, of central concern to this book is sleep's role in memory consolidation. As introduced above, system consolidation is one half of the *Complementary Learning Systems* model, in which the hippocampus captures widely distributed patterns of activations to represent events and then participates in the slow integration of those patterns into cortical memory systems during sleep. The standard model at present is *active systems consolidation*:

> Active systems consolidation accounts for long-term memory formation in the hippocampus-dependent episodic memory system, with a particular focus on the role of sleep.... During sleep, the

---

[45] Deboer, "Sleep homeostasis and the circadian clock," p. 70.

[46] Giulio Tononi and Chiara Cirelli, "Sleep and the Price of Plasticity: From Synaptic and Cellular Homeostasis to Memory Consolidation and Integration," *Neuron* 81 (January 8, 2014), p. 12. For a review of the interesting phenomenon of unihemispheric sleep, see Gian Gastone Mascetti, "Unihemispheric sleep and asymmetrical sleep: behavioral, neurophysiological, and functional perspectives," *Nature and Science of Sleep* 8 (2016), pp. 221-38.

hippocampal neuronal representation is repeatedly reactivated, with these reactivations propagating across the associated memory network and simultaneously reactivating its distributed neocortical components. Synaptic consolidation processes triggered by this coactivation strengthen the memory representations in the neocortex and thus allow the slowly learning neocortex to integrate them into pre-existing long-term memories. Importantly, the systems consolidation process during sleep does not simply increase the long-term stability of memories. Because it reorganizes the memory, with the neocortex integrating many related and overlapping experiences, consolidation also yields representations at higher levels of abstraction, generalization, and efficiency.[47]

According to the model, coordinated waves of burst oscillations in the hippocampus effect this integration of patterns through a combination of replay to strengthen synaptic connections representing significant new experiences and of downscaling (decreasing the strength) of synaptic changes tagged as mere noise.

The oscillations in slow-wave sleep in stages 2 and 3 of NREM are about 1 Hz. During this cycle, the depolarized, spiking component part of the wave is called the up-state, when the neurons are active and can change their synaptic strength. The quiet, hyperpolarized half of the wave is called the down-state. The thalamus generates sleep spindles (short bursts of 12-15 Hz waves), but these spindles seem to be more localized to regions of the neocortex that participated in learning during wakefulness. These spindles propagate to the hippocampus, in which interactions between the recurrent network and inhibitory interneurons of CA3 produce sharp-wave ripples that replay the neuronal patterns of activetion recorded both in the hippocampus itself and in the connection between the hippocampus and affected cortical regions.[48] The spindles during the up-state

---

[47] Klinzing et al., "Mechanisms of systems memory consolidation during sleep," p. 1598. This review provides an excellent overview of the topic, even if it does not explore some of the controversies being debated in the field, which I subsequently touch upon.

[48] Hannah R. Joo and Loren M. Frank, "The hippocampal sharp wave–ripple in memory retrieval for immediate use and consolidation" *Nature Reviews Neuroscience* 19 (December 2018), pp. 744-57. Initially, it was believed that activations started in the hippocampus and

involving the neurons participating in the event representations are believed to enhance those patterns of connectivity and also to mark them for protection from either generalized or specifically targeted downscaling.[49] The role of REM sleep after slow-wave sleep is still under investigation, but REM sleep also appears to induce downscaling through selective pruning of synapses not marked for preservation during slow-wave sleep.

Many aspects of the processes that implement memory consolidation remain uncertain. Tononi and Cirelli argue that consolidation primarily proceeds by down-selection of synapses. The argument about whether episodic memories ever become fully corticalized rather than relying on the hippocampus to serve as an indexical network to bind the components together remains as strong now as it was twenty years ago. However, the goal of this chapter is to clarify the neurobiological requirements for synthesizing the broadly distributed patterns of activation into episodic and semantic memory that together comprise the high-order internal model of the self and the world. These patterns arising from experience include raw sensory data, attentional data, affective assessment, comparisons with previous memories, with current contexts, and with plans for possible action. While the disagreements about details of the biology underlying this synthesis undeniably are important for clarifying the processes that drive memory, there remains a broad consensus about the nature of the construction of memory that is central to my inquiry. Researchers agree that the important patterns encountered in events and represented in episodic memory do in fact

---

then spread outward to the relevant cortical regions, but new research has shown a reciprocity or looping between cortical regions and the hippocampus. It turns out that hippocampal "replay" of patterns of activation are not very accurate, and it has been suggested that the activated patterns reflect not single events but a statistical averaging of similar events represented in the hippocampus.

[49] This downscaling is crucial for proper management of the neuronal networks, and some of it occurs during slow-wave sleep. Tononi and Cirelli in their *synaptic homeostasis hypothesis* strongly argue that downscaling during slow-wave sleep is in fact the core mechanism behind memory consolidation. They present their updated model in Giulio Tononi and Chiara Cirelli, "Sleep and synaptic down-selection," *European Journal of Neuroscience* 2020 (51), pp. 413-21. For a more mainstream account, see, for example, Niels Niethard and Jan Born, "Back to baseline: sleep recalibrates synapses," *Nature Neuroscience* 22 (February 2019), pp. 149-53.

become integrated into semantic structures and that this integration broadens the semanticized and complexly contextualized elements of episodic memory that participate in recall.

The basic model for the consolidation of memory remains the Complementary Learning System (CLS): fast encoding of the distributed patterns of neuronal activations that represent events, followed by a slow integration. Researchers in cognitive neuroscience have studied the nature of the processes supporting the consolidation of memory, through particular tasks that focus on distinctive features of a wide range of neuronal systems.[50] They use these tasks to examine the systems' speed of acquisition, their susceptibility to interference, the contributions of REM and NREM sleep, and speed of extinction as well as the regions in the brain that are most involved. Throughout these experiments, the CLS model holds up reasonably well. Replay of increasingly processed patterns that coordinates the hippocampus and relevant regions entrains the cortical networks to accommodate the new patterns (Figure 6.9).

This system of processes, then, supports the *encoding* of memory. What happens in *recall?* This question has two components. The first is *what* is recalled in recalling a memory; the second is *how* recall is orchestrated. Neuroscientists know a good deal about the first question, especially in relation to the patterns of activity that were part of the initial encoding. The patterns for recalling an event differ from those activated on first experiencing that event. In the initial encoding the trisynaptic pathway in the hippocampus effects a profound restructuring of the relationship of the details of events to their contexts, a restructuring which makes the ability to access memories of the particulars far more flexible in a way that deeply changes how humans engage experience. For example, it has been argued that the trisynaptic pathway in the hippocampus is

---

[50] Motor learning proceeds by mechanisms independent of the MTL memory system, but movement tasks that require learning place patterns fall back into the range of problems handled by the MTL system.

not well developed in young infants, who must rely on just the monosynaptic pathway from the entorhinal cortex to CA1:[51]

> Infants as young as 6-months old can imitate actions the day after the demonstration..., but their memory is highly specific in that they fail to imitate when features of the puppet (e.g., color or shape) or the location in which the task is performed is changed.... Conversely, generalization in deferred imitation is first observed at 12 months when infants can retrieve action sequences using puppets that differ in color from the demonstration puppet, and by 18-months, infants can generalize their memory across puppets that differ in color and shape and across novel environments (e.g., imitation in the laboratory when demonstrated at home).[52]

Similarly, patients with anterograde amnesia can be taught new items of semantic memory through repetition (i.e., statistical learning), but their ability to draw on these items suffers from the same constraints experienced by young infants, and their recall is limited to only the context in which they learned the item.[53] Thus, the flexibility in recall of both episodic and semantic memory appears to derive from the complex interactions of pattern separation and pattern completion effected in the trisynaptic pathway of the hippocampus. The hippocampus restructures the relationships between the elements in the representation of an event—and learning a new word is such an event—allowing the patterns of connectivity that represent one element to activate the broader patterns of connectivity that bind all the elements. We take this capacity for granted, but it

---

[51] For a recent review, see Ellis et al., "Evidence of hippocampal learning in human infants."

[52] Adam I. Ramsaran, Margaret L. Schlichting, Paul W. Frankland, "The ontogeny of memory persistence and specificity," *Developmental Cognitive Neuroscience* 36 (2019), p. 6. Also see Sinéad L. Mullally, "Commentary: Elucidating the neural correlates of early childhood memory," *International Journal of Behavioral Development* 39.4 (2015), pp. 306-07, and Rebecca L. Gómez and Jamie O. Edgin, "The extended trajectory of hippocampal development: Implications for early memory development and disorder," *Developmental Cognitive Neuroscience* 18 (2016), pp. 57–69.

[53] Stark and Stark, "Introduction to Memory," p. 845.

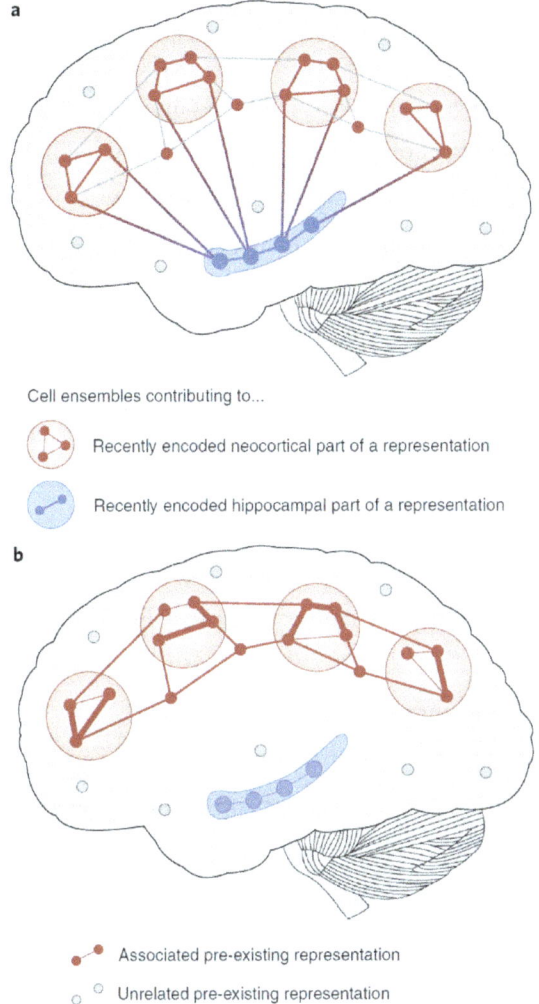

Figure 6.8: Memory Consolidation[54]
(a) The red nodes in the cortex and the blue nodes in the hippocampus connected by heavy read lines (synaptic pathways) participate in the newly learned representation of events. The lighter blue pathways between cortical nodes represent preexisting patterns of synaptic connectivity. **(b)** After repeated activation of the new synaptic pathways, the cortical networks change to accommodate the new patterns of neuronal connectivity. This model illustrates the corticalization of memory but is agnostic about whether the new networks include both episodic and semantic memory or just semantic memory.

---

[54] Klinzing et al., "Mechanisms of systems memory consolidation during sleep," p. 1600.

is a remarkable neurobiological feat. The restructuring of patterns that allows the recall of memories across changes in context is the most striking aspect of the *what*—the changes in content—in recalling memories.

The second aspect of changes in the content of memory is less radical: it does not need the specialized hardware of the dentate gyrus and CA3, and instead is a consequence of the integration of the elements of episodic memory into larger semantic structures. In the experiences represented in episodic memory, the objects and events (as categories of action) are not unitary "things" but patterns of activation that draw together sensory, affective, and cognitive information. However, the initial patterns that integrate and record the elements of unique events are still narrowly connected within the structures assembled through the hippocampus. Consolidation makes these connections broader, deeper and more complex, so that, when recalled, the reassembled activations retain and reflect these broader connections. Recollection contains more information than encoding.[55]

Although memory consolidation integrates new widely-distributed patterns of *activity* into patterns of *connectivity*, researchers have been exploring the process by which this consolidation happens. It is clear that this is a gradual process, but how gradual? Many studies have demonstrated improved performance on cognitive tasks after sleep, with a range of days over which improvement can be tracked. While testing specifically for effects produced by the integration of specific experiential elements into larger semantic frameworks remains something of a challenge, one recent technical development that has proved powerful in studying the difference between the patterns of activation that participate in encoding and those activated by recall is *multivoxel pattern analysis* (MVPA), which allows researchers mathematically to compare sets of fMRI images that are otherwise difficult to analyze. Since the base data is fMRI imagery, the many brain regions that participate in the activations can be analyzed together to identify clusterings of similarity and difference. One result

---

[55] See, for example, Serra E. Favila, Hongmi Lee, and Brice A. Kuhl, "Transforming the Concept of Memory Reactivation," *Trends in Neuroscience* 43.12 (December 2020), pp. 939-50.

of this sort of analysis is that, not surprisingly, there is greater transformation of patterns of activation in the higher-order association areas than in lower-level perceptual areas.[56] The conclusion is worth stressing:

> Whereas the abstraction processes enable the formation of conceptual knowledge from perceptual experience to support highly complex functions, such as language, creative thinking, and problem solving…, the construction processes allow the flexible use of past information to serve current and future goals.[57]

I so far have not discussed *how* exactly the brain manages the reactivation of the patterns of connectivity that represent consolidated memories. This is because the problem is difficult to investigate. Given the nature of the tools currently available to explore human brain activity, we still lack the temporal and spatial precision to track the crucial dynamics of reactivation. It remains possible, however, to discuss some aspects of the phenomenon. We know, for example, that, in the initial encoding of patterns representing experiences, the MTL system draws on patterns stored not just in CA3 but also in cortical regions. We know, moreover, that the attentional neuronal networks that assess perceptual data draw on stored patterns. More generally, since the brain is a predictive processor at all levels, it relies extensively on access to the vast model of the world and the self synthesized in memory in the moment-by-moment unfolding of experience.

---

[56] Xiaoqian Xiao et al. note:

> Whereas the representation in the IPL [inferior parietal lobule] is abstract and identity-specific, invariant to viewpoint and other features, the VVC [ventral visual cortex] representation contains perceptual details…. Consistently, it is proposed that the IPL lies at the convergence of multiple perceptual processing streams, enabling the progressive abstraction of conceptual knowledge from perceptual experience…. The representation in the IPL, but not the visual cortex, is modulated by semantic similarity.

Xiaoqian Xiao, Qi Dong, Jiahong Gao, Weiwei Men, Russell A. Poldrack, and Gui Xue, "Transformed Neural Pattern Reinstatement during Episodic Memory Retrieval" *Journal of Neuroscience* 37.11(March 15, 2017), p. 2996.

[57] Xiao et al., "Transformed Neural Pattern Reinstatement during Episodic Memory Retrieval," p. 2996.

"Play it again, Sam"

A fascinating, oddly reassuring experiment designed to examine the differences between encoding and retrieval for various parts of the brain is a good illustration of the state of our knowledge. Researchers took fMRI images of participants watching a 50-minute movie and again as they provided detailed accounts of the movie they had just watched. What was being measured was not the transformations due to consolidation but simply the transformations enacted by the initial process of encoding to recast the activations into patterns appropriate for long-term memory. Using MVPA to calculate degrees of similarity in activation across the brain, researchers discovered both that individual episodes could be distinguished from one another *and* that many of the activations for episodes in the high-order areas were more similar *between people* than they were when compared with activations for other episodes for the same person.[58]

More particularly, the reactivated patterns in "a large set of high-order multimodal cortical areas, including DMN [default mode network] areas, high-level visual areas in the ventral temporal cortex, and intraparietal sulcus" showed that the high-level activations between perception and recall not only were modified, but that they were modified in similar ways among individuals, and the degree of modifications correlated with the memorability of the scenes in the movie.[59] The very fact that the DMN participated in the processes of both perception and recall provides another indication of the breadth of high-order regions for assessment of the self and the world that contribute to the patterns of activations that constitute episodic memories. At the same time, there was little

---

[58] Chen et al. explain:

> These results reveal the existence of a common spatial organization for memories in high-level cortical areas, where encoded information is largely abstracted beyond sensory constraints, and that neural patterns during perception are altered systematically across people into shared memory representations for real-life events.

Janice Chen, Yuan Chang Leong, Christopher J Honey, Chung H Yong, Kenneth A Norman & Uri Hasson, "Shared memories reveal shared structure in neural activity across individuals," *Nature Neuroscience* 20.1 (January 2017), p. 115.

[59] Chen et al., "Shared memories reveal shared structure in neural activity across individuals," p. 123.

modification between perception and recall in the low-level areas for visual and linguistic processing: they do not require synthesis in encoding and recall.

What then are memories? The word really refers to two different "things." Memories as *consolidated* into the neuronal networks that gradually build a model of the world and the self are widely distributed patterns of strengthened synaptic connections that are structurally integrated into and contribute to the evolving model. Memories as *experienced*, either consciously or through the merely potentially conscious recall in the DMN, are a *process*, a reactivation of some or perhaps all of the synaptic patterns woven together to represent the remembered object, place, or event as they have come to participate in the larger world of meaning. Some experienced memories are vague and generic and extend only to higher-order association areas. Some, in contrast, are vivid and reactivate patterns of connectivity in the primary sensory cortices. The totality of memory is the entire world of meaning structured through the complex extraction of patterns of significance in experience. It is the brain's effort, starting before birth, to understand and model the body's encounter with the world.

Semantic Memory and the Construction of Meaning

So far, we have considered how the neuronal networks that encode episodic memory represent our experience and how the components of episodic memories—sensory states, perceived objects, cognitive and affective assessments, and sequences of events—become integrated into semantic memory. It should be clear that semantic memory is extraordinarily capacious, heterogeneous, and distributed in networks of connectivity that reach every cortical region that participates in perception, affect, the planning of action and movement, and all of the unimodal and multimodal association areas.

Specific semantic memories may include synaptic connections to the primary visual and auditory cortices, but the memories—thought of as patterns of neuronal connectivity that produce stable attractor states in some high dimensional representation space—surely are not *in* the primary sensory cortices. Where, then, are these neuronal networks? There remains much debate about this, and a range of regions has been proposed. The angular gyrus, fusiform gyrus, and the anterior temporal lobe, for example, all have been implicated, and surely the core neuronal systems are in these sorts of association cortices. For the purposes of this book, however, having some idea of *how* semantic memories develop from statistical learning and from the consolidation of episodic memories and *what* these semantic memories are as neuronal networks is more important—at least for the moment—than knowing *where* they are. It may be best to return to Damasio's schematic account of convergence-divergence zones for a general way to think about semantic memories. Damasio stresses two aspects of memories. First, they are in *dispositional* form: many authors have noted that the point of the brain is to guide action, and memories have valences and connections to plans that serve to inform action. Secondly, memories are *indexical*: their significance is reconstructed through the reactivation of the network of synaptic connections that hold the elements of the memory together.[60]

We can divide semantic memories into four interconnected types: objects, events, lexical semantic memories (i.e., words), and combinatorial (mostly sentential) semantic memories. The four types all blend into one another's neuronal networks. Over the past few days, our cats have been chasing the flies that got past our screens. For me, then, the network that represents "fly" as an *object* is connected to and activates the *event* sequences of my cats leaping after the flies. "Fly," also evokes a much older personal memory: the sentence "En boca cerrada no entran moscas" ("Flies do not enter a closed mouth"), a Spanish saying taught by my twelfth-grade Spanish teacher, Señor Gomez. "Fly," a lexical item, calls forth images of both non-verbal events and memorized sentences: thus, the neuronal networks encoding objects can include all the other types of memory structures, and it makes little sense to consider them simpler than the others. Still, objects are the easiest to grasp. Their role in pattern completion in perception, for example, is fairly direct: a ball partially hidden by a chair leg gives our perceptual system enough detail to identify it and to fill in our awareness that the unseen part of this known object is there. Given this fundamental role, we can think of "objects" as the prototypical attractor basins in high dimensional representation spaces defined by the matrix logic of neuronal networks. Since

---

[60] Damasio *Self Comes to Mind*, pp. 143-61.

"events" entail sequences of action and unfold through space as well as time, conceiving of their encoding is more complex than that of objects. Still, there is a sense in which "events" are fixed structures—"things"—even if they extend in time and space, often participate in cause-effect structures, and may connect to networks of intentions and goals. "Events," as learned neuronal memory structures, bind these elements together so that, in their complex coordination, they can be reactivated on recall.

Words behave differently. They combine differently. Words are phonological or orthographic indices that connect to lexical semantic structures. The vast networks of lexical semantic neuronal structures are activated to participate in linguistic usage, a very distinctive form of combinatorial semantic organization that requires that lexical semantic networks possess special features to enable the free-flowing, flexible, but still constrained combinatorial structuring of language. However, how do words, as units of combination, signify? What is the scope of the neuronal networks that represent words in the brain? Before turning to the problem of how the brain facilitates the combination of activations of lexical networks to both create and understand streams of language, we need to briefly consider the *embodiment* debate in neuroscientific discussions of the extent of the networks that represent words as lexical semantic objects through their synaptic connectivity.

Neuronal Networks for Lexical Semantic Objects

Do words, in the end, derive their meaning from bodily experience? In a non-trivial sense, they *must* derive from bodily experience, since all the learned structures of the brain are necessarily mediated by the body. However, the intense debates among researchers and theorists about *embodied* (or *grounded*) semantics over the past two decades have had a narrower focus on the role of sensation and bodily action in the meaning of words. When I think of "grasping an idea," do the neuronal patterns of activation that structure the meaning of "grasp" extend to the supplementary motor area, which is responsible for planning complex muscular movements?

George Lakoff and Mark Johnson famously proposed the idea of *conceptual metaphors* in their book *Metaphors We Live By* in 1980. They argued that we have indeed developed a metaphorical usage of the physical action in "grasping" to extend to the action of having a firm, if figurative "hold" on an idea. By a very different route, in 1990, Stefan Harnad, reflecting on the problem inherent in symbolic accounts of semantic meaning and the implications of the emerging field of connectionist modeling, gave a more formal account of "the symbol grounding problem" in cognitive psychology that has proved extremely influential in discussions of embodied meaning in the brain. Harnad asked the questions

> How can the semantic interpretation of a formal symbol system be made *intrinsic* to the system, rather than just parasitic on the meanings in our heads? How can the meanings of meaningless symbol tokens, manipulated solely on the basis of the (arbitrary) shapes be grounded in anything but other meaningless symbols?[61]

That is, individual words are part of a system of words and are defined through other words. How, then do they point beyond this internal system to a world outside the system itself? Harnad proposed that a connectionist model of "dynamic patterns of activity in a multilayered network of nodes" that extracts regularities from sensory input could complement the cognitivist models of the mind as a symbol-manipulating system and provide the needed "grounded elementary symbols" that could then be articulated and combined by symbolic processes in cognition.[62]

Thirty years after the publication of Harnad's paper, debates about symbolic versus non-symbolic (connectionist) representation still continue in cognitive science and philosophy of mind. In neuroscience, however, the issue in lexical semantics has been reframed as a question of *abstraction*: does the brain continue to rely crucially on reactivating synaptic connections to the primary

---

[61] Stevan Harnad, "The Symbol Grounding Problem," *Physica D: Nonlinear Phenomena* 42.1-3 (June 1990), p. 335. [Emphasis in the original]

[62] Harnad, "The Symbol Grounding Problem," pp. 337 and 345.

sensory and motor cortices to represent the meanings of words? Or does the process of consolidating the network of semantic memory abstract the connections representing words away from low-level networks and, instead, construct meaning through the word's relation to all the other words neuronally represented in the higher-order system? Or is there a middle ground? Does the representation of words as lexical items preserve the synaptic connections to low-level cortices but also create networks of relations to other lexical nodes in the system?

In support of embodied semantics, it is clear that the reactivation of sensory cortices can be a significant component of processing written and spoken words. Indeed, all the models for memory consolidation discussed above propose that synaptic connections that extend to the primary sensory cortices are part of both the episodic memory engram and the semantic memory for the objects and events. That is, semantic memory in the standard model is inherently embodied. Still, the question remains of how the networks of synaptic connections for the objects and events get bound to words, the phonological and orthographic tokens. An increasingly nuanced understanding of the regions of the brain that participate in processing lexical semantics has led some researchers to propose a range of models of how lexical items are integrated into specifically multimodal networks. In these models, the network for lexical meaning draws on all the synaptic connections that encode semantic memories. These include not only unimodal sensory and association cortices, but also the motor planning cortices and the networks that represent the interoceptive and affective responses associated with experience.[63] Since a role for the primary sensory and motor cortices is implicit in contemporary models for memory consolidation, the persistence of the debate suggests that there is more than biology at stake in the question of embodied semantics. And indeed, stressing that semantics is embodied takes on

---

[63] See, for example, Max Garagnani, Evgeniya Kirilina, and Friedemann Pulvermüller, "Semantic Grounding of Novel Spoken Words in the Primary Visual Cortex," *Frontiers of Human Neuroscience: Speech and Language,* February 24, 2021, doi: 10.3389/fnhum.2021.581847; and Karalyn Patterson and Matthew A. Lambon Ralph, "The Hub-and-Spoke Hypothesis of Semantic Memory" in Hickok and Small, eds., *Neurobiology of Language*, pp. 765-776.

particular significance because of the long rationalist and idealist traditions of separating cognition from the contours and constraints of bodily experience. Research like Lawrence W. Barsalou's work on "situated cognition" (or "grounded cognition") and Friedemann Pulvermüller's explorations of the biology supporting grounded semantics thus not only makes an important contribution to our understanding of the bodily basis for lexical semantics, it also requires that we restructure old traditions of thinking about cognition, one of the most crucial impacts of the development of neuroscience.[64]

Given the difference in conceptual framework, cognitivist paradigms for abstract symbolic processing play little role in the contemporary neuroscience of lexical semantics. However, language as an abstract system of tokens of a very particular sort nonetheless continues to serve as an important empirically-grounded model in neuroscientific approaches to linguistic meaning. Early twentieth century structuralist arguments about the nature of linguistic meaning, derived from both theory and observation, have proven enduring, even as they have been reformulated into a new analytic framework. Words as orthographic and phonological tokens do indeed participate in a vast system of mutual reference and mutual definition that can appear to be detached from any sources of embodied reference. While this sort of model of a self-structuring system of arbitrary tokens that obtain their meaning through mutual differentiation has been part of structuralist thinking about language since Saussure, the current embodiment of a self-structuring system is the paradigm of *distributional semantics* that originated from computational linguistics, and which has proven extremely effective in explaining patterns of brain activity in semantic interpretation (See Box 1).

The discussion of memory in this chapter has stressed that the consolidation of memory by its very nature extracts statistical regularities in experience. The consolidation of memory for *words* as "things" behaves the same way as for other objects experienced in the environment, and one of the greatest

---

[64] See, for example, Lawrence W. Barsalou, "Challenges and Opportunities for Grounding Cognition," *Journal of Cognition*, 3.1 (2020), pp. 1–24; and Garagnani et al., "Semantic Grounding of Novel Spoken Words in the Primary Visual Cortex."

> **Box 1: Distributional Semantics**
>
> Distributional semantics is a way of describing the meaning of words as a function of the patterns of their usage—their distribution—in a domain of language use. Treating words as just arbitrary tokens in a domain, distributional semantics uses mathematical methods that identify regularities of word usage. In the simplest version, the *co-occurrence* for every word in the collection of texts that define the domain with every other word within a specified distance is recorded. If there are 10,000 words in the collection, then the usage of each word can be thought of as a 10,000-element vector (the value of each element is the number of times it co-occurs with the other words). The *degree of similarity* between two words is then calculated by multiplying the two vectors (and normalizing the results): words that have identical patterns of co-occurrence are identical from the perspective of usage. This comparison of usage patterns in the collection produces a form of "meaning" that is based strictly on relations among all the tokens in the collection.

regularities in the experience of words is simply their usage in phrases and sentences either heard or read. While the regularities in usage extracted in the brain surely are more complex than the patterns of co-occurrence used in the simplest version of distributional semantics, both modes of structuring meaning—the biological and the computational—rely on the basic idea of identifying statistical regularities. So, it should not be a surprise that similarity analyses for semantic processing of a set of words in the brain using MVPA or other techniques reveal significant matches with results drawn from a distributional semantic analysis of a large corpus of contemporary texts like Wikipedia, etc.[65] Still, words as tokens in texts are only part—though a large part—of the experiences with language from which we learn meaning. Thus, while distributional semantics works within a closed system of words-as-tokens completely outside any forms of meaning through reference to non-words, in fact much of

---

[65] See, for example, Francesca Carota, Nikolaus Kriegeskorte, Hamed Nili and Friedemann Pulvermüller, "Representational Similarity Mapping of Distributional Semantics in Left Inferior Frontal, Middle Temporal, and Motor Cortex," *Cerebral Cortex* 27 (January 2017), pp. 294-309.

the neuroscientific research using distributional semantic data stresses the *complementarity* of embodied and distributional sources of meaning. A recent paper argues, for example:

> Proponents of grounded or embodied semantics… would argue that the dimensions of representation of distributed models lack a biological grounding. However, these two approaches may in fact be surprisingly complementary.… [C]ombining statistical information about word context with the rich understanding of the physical world typical of human learners may be essential for improving the performance of distributional models to reach human-equivalent levels. For this reason, we prefer to see current empiricist distributional models… not as an alternative to grounded theories of semantics. Instead, we think that they supplement one another, and integrating such perspectives could be a key step in furthering our understanding of human semantic knowledge.[66]

The neuroscientific understanding of words as elements of semantic memory appears to have reached an elegant synthesis of distributional and embodied models for statistically derived meaning represented as widely distributed patterns of synaptic connectivity. There is no need for two separate tracks for learning here: each is simply a different aspect of experience with language that is represented through the human system for the development of semantic memory.

The Expansion of Meaning through Combinatorial (Sentential) Semantics

The phonological and orthographical markers through which words are identified as elements in semantic memory are different from sounds like the bleating of sheep or visual objects like rubber balls. What makes words words

---

[66] Jona Sassenhagen and Christian J. Fiebach, "Traces of Meaning Itself: Encoding Distributional Word Vectors in Brain Activity," *Neurobiology of Language* 1.1 (January 2020), p. 72.

rather than imitations of environmental sounds? In this section I touch only lightly on decades of research in linguistics, in part because linguistics is a huge field with intense debates, but in greater measure because much that is central to linguistics—the actual structures through which we construct and interpret sentences—can be bracketed for the purposes of this study as a level of detail concerning the specific results of neuronal processing that is less important than the organization of the networks that effect this processing. The linguistic account of these processes of construction certainly will add important elements to our discussion, but for present purposes, a bare-bones presentation of the neuronal processes that support compositionality (i.e., the construction and interpretation of phrases, sentences, paragraphs, poems, and books) will suffice. This simplified account of what makes words words and allows them to be combined into larger strings with complex semantics begins, then, with a semantic feature shared by all words that serves as an elemental basis for composition: *thematic roles* that are either assignable *to words* (mostly nouns) or required *by them* (mostly verbs) in phrase construction. To be a bit more precise, linguists distinguish between *thematic relations*, a semantic term, and thematic roles, or *theta-roles*, which are part of syntax.[67] Common thematic relations/roles are *agent* (the doer of actions), *patient* (the recipient of the action), *instrument* (the means of effecting an action), and *location* (where the action occurs). The overlap between the two terms underscores their role at the semantic-syntactic interface, the transformational process by which patterns of associations constructed from semantic memory attain syntactically regular form and by which, conversely, sentences structured through syntactic rules become compositional structures of neuronally activated semantic meaning.

---

[67] The manner in which thematic relations are represented in the neuronal structures of semantic memory is unknown. However, a recent article suggests that thematic relations are a part of "core knowledge" about the world (available pre-linguistically) that also come to serve a structural role in the delimited linguistic system. See Lilia Rissman and Asifa Majid, "Thematic roles: Core knowledge or linguistic construct?" *Psychonomic Bulletin & Review* 26 (2019), pp. 1850–1869.

The Neuroscience of Language Processing

Projects to study the neuroscience of language processing confront the challenge of not having non-human primate or rat analogs. Animal models allow researchers to study the micro-level neuronal interactions that innervate brain processes. For speech and reading, the only tool that scientists initially had were brain lesion studies and occasional open-brain probes during surgery. Through these tools, biologists identified Wernicke's area in the superior temporal gyrus and Broca's area in the inferior frontal gyrus, which, when damaged, produce two distinct types of aphasias. In Wernicke's (fluent) aphasia, patients spoke smoothly but strung together nonsense syllables or words that made little sense in context; in contrast, in Broca's aphasia, patients had trouble speaking words at all.[68] In the century since Wernicke's studies, neurologists have carefully differentiated a wide variety of aphasias for speech production and comprehension associated with damage to brain regions near either Broca's or Wernicke's area. However, it has been only since the advent of fMRI and other forms of neuroimaging that neuroscientists have been able to get fine-grained information about both aphasiac and normal speech processing. As it turns out, Broca's and Wernicke's areas remain very much part of the story, although other cortical areas also play crucial roles, as do a set of white matter tracks connecting posterior and anterior cortical regions.

One widely used account of language processing proposes a dual stream (dorsal and ventral) analogous to the two-stream model of visual processing. The first step (the green box in Figure 6.10) is converting the stream of auditory cortical output into phonological data. That data then goes to both an "auditory-motor" dorsal stream involved in the motor control needed for speech production and to an "auditory-conceptual" ventral stream for structuring and interpreting the content of speech.[69] The organization of the feedback loop between

---

[68] Pierre Paul Broca identified the region named after him in a study of autopsies of aphasiac patients in 1865. Similarly, Carl Wernicke proposed in 1874 that damage to a region in the superior temporal sulcus produced forms of fluent aphasia.

[69] Gregory Hickok and David Poeppel, "Neural Basis of Speech Perception," *Neurobiology of Language*, p. 299.

hearing the spoken words and adjusting speech production in the proposed dorsal stream provides another example of the pervasive functioning of predictive processing in the brain. The articulatory motor-control network predicts how speech will sound, while the auditory networks provide the feedback to correct the speech production as needed.[70] The two primary regions of the dorsal articulatory system—the phonological mapping component in the superior temporal gyrus and the articulatory component in the posterior inferior frontal gyrus—are connected by two parallel white matter tracts, the arcuate fasciculus and a portion of the superior longitudinal fasciculus.

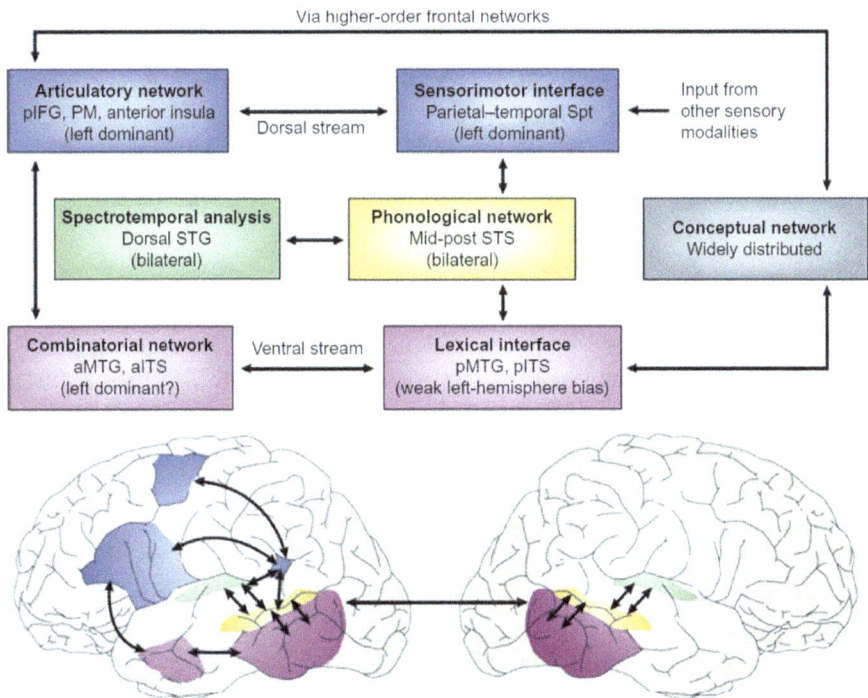

Figure 6.9 Neural Networks to Support Language Perception and Production[71]

---

[70] Gregory Hickok and David Poeppel, "Neural Basis of Speech Perception," *Neurobiology of Language*, p. 303.

[71] Gregory Hickok and David Poeppel, "Neural Basis of Speech Perception," in Gregory Hickok and Steven L. Small, eds., *Neurobiology of Language* (London: Academic Press, 2015), p. 300.

The ventral language pathway maps phonological data to words as networks in semantic memory, finds links to connect small units of word-patterns to phrases that serve thematic roles, and then, from the small units, constructs the complex syntactic structures of speech and discourse. There are debates about how and where these operations are performed. Several accounts identify the first phase, moving from words as semantic patterns to words bound to thematic roles, with the anterior temporal pole (ATL), which also has been proposed as the hub in the hub-and-spoke account of semantic memory.[72] A recent study offers a good way to describe the distinct processes behind the phrase construction from semantic information in the ATL and the next phase of building more complex syntax. In linguistics, there are two competing approaches to parsing the syntax of sentences.[73] The first, dependency grammar, looks for dependency relationships between the words of a sentence. In this mode of analysis, for example, we have:

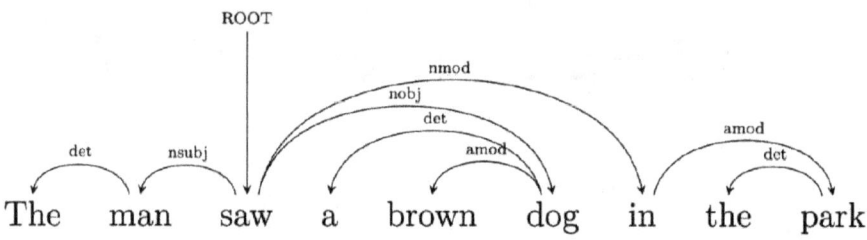

Figure 6.10: Dependency Parsing[74]

"The" depends on "man" as a determiner; "man" depends on "saw" (which, as the verb, is the root of the sentence) as the subject; "dog" depends on "saw" as the

---

[72] Hickok and Poeppel observe, "ATL regions have been implicated both in lexical-semantic and sentence-level processing (syntactic and/or semantic integration processes) …. Higher-level syntactic and compositional semantic processing might involve the ATL" (p. 303) Ryan Staples and William W. Graves, "Neural Components of Reading Revealed by Distributed and Symbolic Computational Models," *Neurobiology of Language* 1.4 (September 2020), p. 395. For the "hub-and-spoke" model, see Patterson and Lambon Ralph, "The Hub-and-Spoke Hypothesis of Semantic Memory."

[73] Alessandro Lopopolo, Antal van den Bosch, Karl-Magnus Petersson, and Roel M. Willems, "Distinguishing Syntactic Operations in the Brain: Dependency and Phrase-Structure Parsing," *Neurobiology of Language* 2.1 (January 2021), pp. 152-75.

[74] Lopopolo et al., "Distinguishing Syntactic Operations in the Brain," p.157.

object, and so on. One can think of the first phase of semantic-syntactic processing as the creation of the set of dependency relations in the sentence, before they all are integrated into a sentence structure.

The second phase of combinatorial processing by most accounts is the production or analysis of higher order syntactic structures that represent correspondingly higher order semantic constructions. The syntactic/semantic object looks something like a phrase-structure grammatical tree:

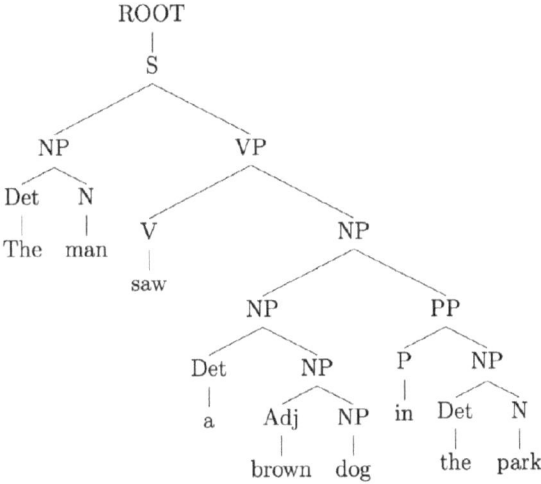

Figure 6.11: Phrase-Structure Parsing[75]

Experiments that probe the brain's response to complexities in syntactic structure suggest the primary sites for the neuronal activation in the synthesis/analysis of these combinatorial structures are in the superior temporal gyrus and the inferior frontal gyrus.[76] The relation between semantic and syntactic organization appears complex yet close: Broca's area (pars opercularis, Brodmann 44) in coordination with the STG, connected by the arcuate fasciculus, appears to do the lower-level semantic/thematic dependency ordering while the pars triangularis (Brodmann 45) in conjunction with portions of the superior and medial temporal gyrus serves as the high-level syntactic/semantic interface. These regions are connected by the inferior fronto-occipital fasciculus (IFOF), where a superficial layer

---

[75] Lopopolo et al., "Distinguishing Syntactic Operations in the Brain," p. 156.

[76] Lopopolo et al., "Distinguishing Syntactic Operations in the Brain," p. 153

supports specifically verbal semantics while a deeper layer supports amodal semantics.

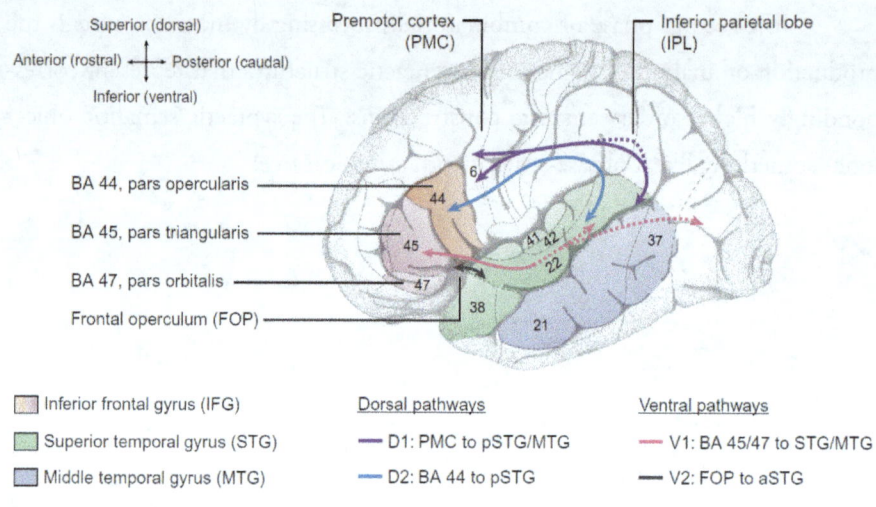

Figure 6.12: Connectivity in the Language System[77]

I note the white matter tracts involved in language processing because the IFOF in particular appears to be distinctively human, with no corresponding structure even in non-human primates. The connectivity between posterior and frontal cortical regions that it provides may be extremely important in explaining the human capacity for complex semantic structures. The standard pattern throughout the brain is that as neuronal networks pass activations to subsequent networks, the dimensional structure of the receiving networks build on the processing of the input networks to grow more complex, and the networks thus construct ever higher-order patterns. The communications between the regions in the ATL, medial and superior temporal gyrus and the inferior frontal gyrus seem to support the transformation of rather simple experiential patterns preserved in semantic memory into greatly expanded combinatorial possibilities for meaning. Language is not just language: the instrumentality of syntactic forms gives extremely flexible structure to the flow of "thoughts" and provides a

---

[77] Angel D. Friederici, "The Neuroanatomical Pathway Model of Language: Syntactic and Semantic Networks," *Neurobiology of Language*, p. 351.

mechanism for ordering, reordering, and recreating all the patterns of experience preserved in our vast network of episodic and semantic memory. Exactly *how* this complex ordering of structure and signification works remains a mystery.[78] Still, we have some knowledge of the neuronal components of this process and have begun to explore the patterns of connectivity that reach far beyond the unit-level phonological and orthographic encoding of "words" to synthesize large-scale semantic structures (represented as sentences, paragraphs, books) that persist in semantic memory and ramify through all the modalities of bodily experience. It remains worth stressing that these combinatorial semantic structures—once generated—then can become consolidated into semantic memory and serve as the basis for yet further construction.

Poems and Identity

What does the evolving neuroscience of language add to our understanding of meaning in human experience at present? If the neuronal self, expansively conceived, is the combination of response systems (perceptual, affective, action) and the totality of memory structures that participate in the assessments that drive response, then all that we remember of linguistic experience (as specially structured combinatorial semantic networks that contribute to the entire semantic matrix) is an integral part of the self. Our linguistic experience is part of how we understand ourselves and the world and part of how we formulate our responses. This has huge significance. The books we have loved, the words we have spat out in anger or used to caress those we love, the narratives of nation, class, and race in which we have been immersed all inform who we are.

---

[78] As Angel D. Friederici sums up the state of neuroscientific knowledge,

> At present, it is not entirely clear how to model these sentence-level semantic processes. What is clear, however, is that the anterior temporal lobe, the inferior frontal cortex, and the posterior temporo-parietal cortex are involved in sentence-level semantic processes, but that their interplay in the service of language understanding remains to be specified.

Friederici, "The Neuroanatomical Pathway Model of Language: Syntactic and Semantic Networks," p. 354.

Let me suggest that what the neuroscientific account of language and semantic memory adds is a nexus of related elements that can change how we think about experience and the self. The most basic is simply embodiment: language is learned and experienced by a biological organism. The first steps of connecting sounds to things and events as words happen within a particular self with a body, dispositions, and prior experiences. Words are learned through successive encounters that create the statistical regularities needed to link specific semantic (perceptual, affective, and cognitive) patterns to sounds. The second aspect, following from the first, is systemic neuronal mediation: words are not defined by simple ostensive reference to the external world: they accumulate within an evolving high-dimensional semantic space. As Lev Vygotsky pointed out long ago, the internal structures are not the same as what we take to be the external structure.[79] The third crucial aspect returns to a different facet of embodiment: the neuroscience of lexical semantics proposes that, given the statistical nature of our learning language, distributional semantics—the social world of language use—reflects much of how we build the internal representational system through relating words to other words. However, at the same time, the patterns of neuronal connectivity that embody words within the matrix of semantic memory maintain their connections to the broader networks that structure the semantic meaning of experience. These include connections to the sensory and affective elements of experience.

And what does the neuroscientific tale of the assimilation into the self of complex combinatorial semantic structures mediated by language signify? Why does it matter? Although I am confident that there are many literary traditions with similar stories, let me draw an example from classical Chinese poetry, which is my area of study. Well-educated Chinese, especially those with literary interests, have memorized hundreds or thousands of poems throughout their lifetimes. To an important extent, these poems inform lived experience. The Mid-Autumn Festival, the fifteenth day of the eighth lunar month, when there is a

---

[79] See Lev Vygotsky, *Mind in Society: the Development of Higher Psychological Processes*, edited by Michael Cole, Vera John-Steiner, Sylvia Scribner, and Ellen Souberman (Cambridge, MA: Harvard University Press, 1978), pp. 80-81.

full moon, for example, evokes poems for the festival written on the theme of seeking to reunite with one's family and leads many people even today to recall in particular the last lines of a song lyric by Su Shi (1037-1101):

人有悲歡離合，	Among people, there are sorrow and joy, parting and reunion;
月有陰晴圓缺，	For the moon, there are clouds and clearing, waxing and waning;
此事古難全。	These matters, from of old are impossible to make whole.
但願人長久，	I only wish that always for us
千里共嬋娟。	Across a thousand miles we can share in this night's charm.

As captured in—and then represented by—this poem, the festival is a moment for thoughts of family, reflection on the vicissitudes of life, and hopes for the future. That is, the mid-autumn moon as it floats in the night sky is no simple object but is bound by this popular poem to thoughts and emotions that are both personal—since every family's circumstances are different— and communal. And these thoughts and emotions—episodically and semantically—have become part of the matrices of the meaning of experience and one's own identity. The moon has become like this. And this moon, with all its confluence of meanings, is part of who we are.

Chapter Seven

Being Biological

Where has this book taken us? The previous chapters have offered neuroscientific accounts of vision, attention, emotion, memory, and the emergent self that break the seeming simplicity and immediacy of each—their experiential givenness—into a dispersed array of components—neuronal subsystems—and reassemble them in ways that vastly complicate their interconnections.

The neuronal networks that I have sketched out in this book are too varied to be fully summarized here, so I will confine myself to the following reprise of the essentials. At the core of the story of interlocking systems that provide the substructure for experience are the affective nuclei of the brainstem and midbrain. They are the beginning, before any cortical elaboration or external input. These affective nuclei have been shaped by evolutionary pressure to give a particular human structure to the earliest forms of the intentionality through which we engage, respond to and learn about the world. From this kernel of low-level signaling, the deeply social self and the world as we apprehend it develop. The midbrain dispositional matrix, building upon its most immediate responses, creates experientially driven secondary and tertiary cortical structures suffused with intentionality. This affective matrix ramifies though our systems for perception, cognition, and memory, and shapes our necessarily embodied understanding of ourselves and the world.

The models I have developed in this book focus on what I believe are the aspects of the evolving neuroscientific framework that are most immediately relevant to humanistic inquiry. To return to the question, what then do these models tell us? They suggest a few significant implications. To start, the neurobiological reconstruction of experience unravels some old, convenient binarisms in our discussions of human experience. Perception, often contrasted to cognition, for example, is not purely sensory in the neuroscientific account but has a cognitive dimension: even simple acts of seeing are informed by the affective and cognitive assessments that drive attention. Affect, moreover, is not the irrational opposite of rational cognition: cognition (usually thought of as being performed by a *res cogitans*) does not somehow float above the body (the *res extensa*) since, among other reasons, affective elements suffuse both the structuring of memory upon which cognition draws as well as the direct functioning of the neuronal networks that implement cognitive processes. The stuff of cognition is affectively informed. And affect as a biological mechanism cannot meaningfully be described as irrational. However, the history of particular experiences that shape the articulation of the structuring of the self is not accessible, so we have no direct insight into the logic of affect within cognitive processes.[1]

Secondly, the neuroscientific account in this book demonstrates the sheer connectedness of human existence. The point of the brain is to act, and the biological foundation of intentionality that grounds our experience of the world intimately and inextricably connects our selves to our bodies, to our past, to one another, and to the built and natural environment around us. All of this we retain in our evolving selves.

Beyond these basic generalizations, what do we do with—how do we understand—this model of a profoundly connected biological self? The goal of this book was to set out basic terms and a framework to help us collectively engage in conversation with contemporary neuroscience and begin to integrate its still-evolving perspectives into humanistic inquiry. This vast, complex project, already well underway in many areas, will occupy the human sciences for decades to come. Yet humanistic inquiry for centuries has been exploring questions of

---

[1] Antonio Gramsci, for instance, writes:

> The starting point of critical elaboration is the consciousness of what one really is, and is 'knowing thyself' as a product of the historical process to date, which has deposited in you an infinity of traces, without leaving an inventory.

Antonio Gramsci, *Selections from the Prison Notebooks*, trans. Quinton Hoare and Geoffrey Nowell-Smith (New York: International Publishers, 1971), p. 324, cited in Cassandra Falke, *The Phenomenology of Love and Reading* (New York: Bloomsbury Press, 2017), p. 163.

experience and meaning based on models like those proposed in this book. The neuroscientific accounts do not present a complete rupture in our understanding of the human but a recasting of certain types of questions to shift our perspective on the dynamics of the construction of meaning. Rather than a simple, all-inclusive, reductive framework, the neuroscientific models instead provide a substructure for our accounts of the patterns of experience. These models in themselves do not explain experience: that is our job. The challenge is to see the connections between neuroscientific levels and our higher-order accounts of experience.

In this chapter I provide two illustrative examples of such connections drawn from my own field of literary studies to suggest the roles that biological models can play in our understanding of human experience. The first example draws on Cassandra Falke's application of Jean-Luc Marion's "erotic phenomenology" to analyze the dynamics of literary reading.[2] As a second example, I illustrate how the classical Chinese poetic tradition is grounded in an understanding of selfhood and the world with strong analogies to—and consistent with—the neuroscientific framework. The model for meaning in experience in the Chinese poetic tradition offers us ways to think about the neuronal self and our lives as biological beings.

Phenomenological /Connections

The systems described in this book—vision, emotion, and memory—are pieces of what could be described as a phenomenology, an account of the manner in which we have access to the things that appear before us. Chapter Three discusses how the retina transmits activations to the primary visual cortex, how the complex feedforward and feedback networks pass this information to ever higher orders of processing that identify objects and the meanings attached to those objects. However, this is *not* phenomenology as it has developed in the Western philosophical tradition, especially since Edmund Husserl. Husserl asks us to consider the moment of apprehension before there are any labels or any

---

[2] Cassandra Falke, *The Phenomenology of Love and Reading* (New York: Bloomsbury Academic Press, 2017). I am grateful to Anna M. Shields for bringing this work to my attention.

conceptual orderings (setting aside what Husserl calls the "natural attitude"), when we have nothing more than the pure appearance. Husserl and the phenomenologists who followed his lead asked how that moment is possible, and whether it has a structure that we can analyze. The distinction between the empirical and philosophical explorations of how phenomena arise is important and helps underscore the limits of the models I have presented in this book. They do not constitute a phenomenology in the more formal sense, but they do provide a phenomenology "of sorts." While the neuroscientific models present forms of knowing that are restricted to the "realm of appearances," that is, to such knowledge as is possible in the phenomenal realm, that realm—without ontological guarantees—is the world in which we live.[3] Lack of certainty notwithstanding, it is a rich, complex world that merits our efforts to understand it as best we can, and the neuroscientific models—with their "sort of" phenomenology—offer a significant contribution to this project.[4]

To probe this limited phenomenology, I consider how the neuroscientific models in this book map onto the phenomenological approach of Jean-Luc Marion. Exploring their intriguing parallels can serve as one route to reflecting on the implications of the account of perception, emotion, memory and selfhood that I have presented.[5] I draw on the discussion of Marion's

---

[3] See, for example, the discussion in Massimiliano Aragona, "Phenomenology, Naturalism, and the Neuroscience," in Giovanni Stanghelline et al., eds., *The Oxford Handbook of Phenomenological Psychopathology* (Oxford: Oxford University Press, 2019), pp. 273-83.

[4] This task of exploring the phenomenal realm is essentially what Immanuel Kant left us: we cannot know the transcendental subject or "things in themselves," but we do have intuitions of order—aesthetic judgments—about what we observe within the manifold of experience.

[5] Shaun Gallagher in "On the possibility of naturalizing phenomenology" suggests a possible division of labor (though I suspect this was not his intended argument) between a naturalized phenomenology and neuroscience:

> It remains the case that phenomenology is an attempt to do justice to first- and second- person experiences, to explicate such experiences in terms of their meaning. As such, phenomenology does not directly address the questions of subpersonal mechanisms or causal factors.

Shaun Gallagher, "On the possibility of naturalizing phenomenology" in Dan Zahavi, ed., *The Oxford Handbook of Contemporary Phenomenology* (Oxford: Oxford University Press, 2012), p. 89.

phenomenology in Cassandra Falke's *The Phenomenology of Love and Reading*, which does not offer an exhaustive account of Marion's positions but instead applies his framework to the experience of reading in a way that, I hope to demonstrate, echoes the neuroscientific models.

At the heart of Marion's phenomenology is the difficult concept of the "erotic reduction."[6] As noted above, in Western phenomenology since Husserl, the goal has been to systematically grasp the structure of the moment of pure perception when phenomena appear to us. For Husserl, the primary tool for reflecting on phenomena—"things that appear"—without any attendant identity is the "phenomenological reduction," where "reduction" means "leading back" [to phenomena in their mere appearance]. Marion keeps Husserl's *epoché*, the "bracketing" of our naturalistic understanding of the objects and events we encounter, but makes the phenomenological reduction even more radical by proposing the primacy of a moment *even before* phenomena manifest themselves, when one gives oneself over to—commits oneself to—the primary event of manifestation itself. Marion describes this radical commitment without content as the "erotic reduction," a commitment based on love. Marion defines this "love" very broadly to accommodate many forms of love that are not—and cannot be—differentiated in the erotic reduction since there are not yet definable objects, just self-manifesting phenomena. This "erotic reduction" is abstract and part of a larger phenomenological discourse, and the neuroscientific corollary in fact may help clarify the concept and its significance. I have pointed to the brainstem and midbrain affective nuclei as powerfully shaping the emergence of cortical structures, including the self. For a newborn, why look at all? And why look in particular at the caregiver's face, well before there is a semantic category for

---

My argument here, however, is not that the "subpersonal mechanisms or causal factors" provided by neuroscience can explain Marion's phenomenology—which is based on a form of transcendental reduction—but that the sort of phenomenology we might derive from the neuroscientific models in this book should look very much like what Marion proposes and that Marion's analyses help us reflect on analogous implications for the biological version.

[6] Marion explores the many aspects of this erotic reduction in *The Erotic Phenomenon*, translated by Stephen E. Lewis (Chicago: University of Chicago Press, 2007).

"face?"[7] Looking and seeing thus are grounded in a conceptless affectivity that creates a fundamental bond between the child and the world before there *is* any experiential world. Marion captures this act in his erotic reduction.

I believe that for altricial mammals like humans, Marion's erotic reduction surprisingly serves as a better model for the grounding of attentionally directed perception than Gibson's affordances or Lisa Feldman Barrett's allostatic mechanisms, which are often drawn upon in the neuroscientific literature. That is, social commitments appear as emergent properties within nascent cortical systems initially driven by the conceptless affectivity of the midbrain nuclei. Within this developmental process, the self emerges in coordination with the coalescing of the "significant other," and this dynamic of interaction and self-definition is inherently socially and affectively committed in a way that Marion's phenomenology thematizes.[8]

However, even beyond the initial stages of development, Marion's erotic reduction is useful for thinking about the neurobiology of selfhood and experience. The erotic reduction is, in essence, an affectively committed intuition of presence in which the "thing" (that is present) and the "self" (to whom the thing is present) both are still indeterminate. Marion introduces the idea of "saturated phenomena," in which we attend to an object that appears before us, but we have an intuition that the object (or event) has meaning beyond both our immediate apprehension and that which drove our attending to the object in the first place

---

[7] The pattern extraction for "face" begins building the broad cortical semantic category of "face" *after* the initial intensive, directed looking. The directed attention precedes any cortical categorization.

[8] That is, both "affordances" and "allostasis" are fundamentally "solipsistic and autonomous," which does not capture the deep-root internalization of the experienced other in the self. Analogously, "Marion recurs initially to the traditional metaphysical approach of the *self*, which, he argues, remains by definition solipsistic and autonomous. In his view, metaphysics and phenomenology concentrated primarily on the subject's most inner self and thus excluded the Other" (Irina Schulzki, "Love at Loss: Jen-Luc Marion's Concept of Erotic Reduction and Paul Thomas Anderson's *Magnolia*," in Irina Hron, ed., *Einheitsdenken nach der Postmoderne. Figuren von Ganzheit, Präsenz und Transzendenz nach der Postmoderne* (Nordhausen: Verlag Traugott Bautz, 2015), p. 149). The erotic reduction better captures human experience even from the evolutionary perspective: our hyper-altricial developmental arc evolved to build in the incentives that produce the specific character of human sociality.

(our intention toward the phenomenon). Saturated phenomena stand in contrast to "common-law" phenomena for which intentions directed toward them discovers significations that match and exhaust our intuitions about meaning.[9] The seemingly abstract formulation conveys the common experience of surprise and raised interest and, although obviously not Marion's intent, offers a higher-order systematically integrated approach to thinking about an important, ongoing role for the brainstem/midbrain nuclei in daily life. The easiest example is their role in curiosity. What drives curiosity? From the neuroscientific perspective, one can cite the dopamine system or Panksepp's SEEKING network of interacting nuclei. Using Marion's framework as a metalanguage, one begins with a fundamental commitment to openly engage, and in this "erotic" engagement, one has the intuition that there is something here that exceeds our understanding, so one metaphorically or literally "pokes" it. The result may prove uninteresting—not much to see here—so one may drop it. However, on occasion, inquiry reveals a "saturated phenomenon," something there that eludes grasping and captures one's attention.[10] This higher-order phenomenological model elegantly accounts for curiosity, which in neuronal terms we can describe as a foundational affective response generated by Panksepp's SEEKING network.

Cassandra Falke's *Phenomenology of Love & Reading* applies Marion's framework to reading literature, a probing into the unknown that is close to curiosity. Simply put, why bother? Of course, many people in fact do not read with interest when the text at hand appears as no more than a "common-law" phenomenon, in which there is no intuition of something not yet exhausted. Falke certainly acknowledges such uneventful textual encounters, but, needless to say, her focus is on *engaged* reading and how to account for its experiential

---

[9] Falke usefully explains the relationship among the three terms: "Mediating between *intention* and *intuition* is *signification*, which is the system of meaning-making that allows intentionality to receive intuition" [emphasis added]. p. 50.

[10] Marion is aware of the similarity of the revealing of a saturated phenomenon to Immanuel Kant's aesthetic judgment, in which one has the intuition that two phenomena belong together under a higher conceptual category but does not know (yet) what that category might be. The value of Marion's approach is that it provides a driver for aesthetic intuitions in the affective commitment of the erotic reduction.

qualities. Her analysis is productive for reflecting on the implications of the neuroscientific models presented in this book since it stresses three core themes. First, she applies Marion's erotic reduction—a conceptless affective commitment to engage the world beyond the body—to the act of reading. Second, Falke's understanding of the phenomenology and semantics of reading is capacious in a way that mirrors the model of broad interconnectivity of emotion, memory, and sensory apprehension set out in this book. And finally, like Marion, Falke sets aside the a priori givenness of the "self," which she sees, instead, as forming in its engagement with the world of experience and evolving in every transformative semantic (affective, experiential, cognitive) encounter. Her "self" captures the profoundly open, socially committed qualities of the neuronal self.

Falke begins by describing the phenomenologist's task of respecting the complexity and depth of everyday experience:

> Phenomenology asks "How is it that experiences actually appear to me?" and "What is the 'me' that perceives?" For example, if I hear church bells pealing across the city, I assume that someone somewhere is ringing them, but all I really know is the experience of the sound. I put that experience together with other phenomena of my past—the unremembered recognition of cause and effect, the sight of the church tower that morning, a conversation wherein I learned that these bells (the bells of York Minster) are still rung by hand. Each of these memories, whether or not I consciously recollect them, also contributes to the phenomenality of hearing Minster bells. In order to describe how we experience an event such as hearing these bells, phenomenologists examine memory, the experience of selfhood, the body through which we receive sound and other sensual phenomena, our engagement with other people, and the processes by which we let experience in or find it blocked by other experiences.[11]

Falke brings this sense of the complexity of the phenomenological "life-world" to be understood to the project of reading at the same time that she respects the textuality of books. For Falke, truly engaged reading requires the mix of affective

---

[11] Falke, *The Phenomenology of Love and Reading*, p. 17.

commitment and letting go that Marion describes as the erotic reduction, and if we can let go, then we encounter a phenomenon that exceeds our initial intentional grasp. We bring to reading all of our past and attempt to engage a text with its particularities of lived detail that come from beyond us and are new in unexpected ways. We know our own experiences of streetscapes, for example, but can we grasp the import of the winter street scene in Harlem with the yam-seller in Ralph Ellison's *Invisible Man*, which, if we read through the erotic reduction, appears as a "saturated phenomenon," offering intuitions of meaning that push us beyond ourselves. Falke describes the challenge thus:

> In order to give a book room to operate on us as an intuition and signification that comes to us from elsewhere, we must try to refrain from imposing concepts on it prior to reading.... Can I allow myself to be altered by an intentionality and signification that arises from elsewhere? Literary texts offer signification autonomous enough to both invite and resist my intentionality. As long as I allow my intentionality to be overcome by what the text gives, then the process of learning how to let the text come forth can renew the insight-rooted knowledge that Husserl privileges.[12]

Falke's explanation of "insight-rooted knowledge" has the same complex, broadly experiential, interconnected, non-propositional features as does newly integrated, structure-changing learning in semantic memory.[13] And for her, as for Marion and the neuroscientific accounts, the experiences through which the self takes shape of necessity come from "out there," since our own internal resources begin so meagerly: "Because of the limitedness of our intentionality, our lives unfold as a series of these saturated phenomena as our encounters with seemingly stable

---

[12] Falke, *The Phenomenology of Love and Reading*, p. 57.

[13] Falke, *The Phenomenology of Love and Reading*, p. 53:
> If a book helps make visible knowledge, that knowledge is not information about ourselves or information about the book or information about the world, at least not knowledge that can be stated propositionally.... The insight-rooted knowledge that literature promotes enables a specific privileged relation to the world rather than conveying a collection of facts about it. It is thus knowledge we inhabit more than acquire.

objects take place in a way that we cannot predict, capture, or fully remember."[14] If we want to think through the structural relationships between affect, memory, and the self implicit in the neuroscientific model, Falke's phenomenology of reading, based on analogous commitments, offers us one compelling framework.

A Poetry of Embodiment

Because of the limitedness of our intentionality, our lives unfold as a series of these saturated phenomena as our encounters with seemingly stable objects take place in a way that we cannot predict, capture, or fully remember 　　　　Cassandra Falke	錦瑟無端五十絃 一絃一柱思華年 莊生曉夢迷蝴蝶 望帝春心託杜鵑 滄海月明珠有淚 藍田日暖玉生煙 此情可待成追憶 只是當時已惘然	A brocade zither, fifty strings to no end. For each string, each peg, thinking about flourishing years. Master Zhuang in a dawn dream strayed as a butterfly. Wang Di's spring heart, entrusted to the cuckoo. Over the dark sea the moon is bright; pearls form tears. On Indigo Field the sun is warm: the jade gives off haze. Can this feeling have waited to become a recalled memory? It's just that, at the time, it was already indistinct.

As biological beings, our affective commitments underlie our emergent self as well as our understanding of experience. But these commitments grow complex, deeply mediated within the growing structures of our knowledge of ourselves and the world around us. As biological beings, moreover, our selves and our affectively informed grasp of the world are embodied and shaped by this body's history of encounter: we are constrained to know the world through sensory experience and through our evolving cortical means to extract patterns from our interactions. How can we imagine experience, its meaning, and its structure in these terms? This is a new world of neurobiological selfhood in which many experiential categories shift, intermingle, and redefine themselves, and thus present an imaginative challenge. It turns out, however, that the premodern

---

[14] Falke, *The Phenomenology of Love and Reading*, p. 120.

Chinese poetic tradition is built upon assumptions about the self, the world, and the language that remarkably mirror the neurobiological framework presented in this book. Chinese poetry accordingly can give us a glimpse of the richness of biologically understood human experience.

A Different Ontology, a Different Phenomenology

Like the Greeks, early Chinese thinkers were well aware of the instability of objects and identities in the phenomenal realm, the world accessible to the human senses. The Greek solution to this mutability offered by Plato and Aristotle was to look beyond the sensible realm to a perfect, enduring world, of which ours was but a shadow. In contrast, the mainstream view that emerged at the end of the period of great philosophical creativity in Warring States China (403-255 BCE) was a form of phenomenology that shifted the focus of questions of identity to the realm of specifically human usage. The early proto-Daoist philosopher Zhuang Zhou (ca. 369-286 BCE) delivered a devastating, brilliantly satirical attack on the human capacity for knowledge and on the meaningfulness of the words and categories humans assign to objects in the phenomenal realm. In response, the Confucian philosopher Xun Kuang (ca. 310-235 BCE), while essentially in agreement with Zhuang Zhou's critique, argued that Zhuang Zhou had missed the point of language and knowledge, which served human rather than cosmic purposes.[15] Heaven (*Tian* 天, literally "sky") created us as humans, with distinctive features that determined our capacity for experience and shaped the type of knowledge we could possess:[16]

---

[15] Master Xun argued that Zhuang Zhou "was blocked by [his concern for] Heaven and did not understand the human [situation]" (蔽於天而不知人). Brook Ziporyn in his *Ironies of Oneness and Difference — Coherence in Early Chinese Thought: Prolegomena to the Study of* Li 理 (Albany: SUNY Press, 2012), discusses Xun Kuang's philosophy as a "non-ironic response" to the "ironic coherence" among phenomena proposed by Zhuang Zhou (pp. 199-220).

In recent years there has been considerable scholarly interest in Xun Kuang. Eric L. Hutton has compiled a selection of essays that can serve as an excellent introduction to the range of key topics explored in contemporary scholarship in English: see his *Dao Companion to the Philosophy of Xunzi* (Dordrecht: Springer, 2016)

[16] *Tian* 天 can be variously translated as "sky," "the heavens," and "Heaven." Some scholars render it simply as an entirely desacralized "nature." Its status in Xun Kuang's thought was a

On what does one rely to make distinctions? One relies on Heaven-given senses. When things are of the same category and same responsive characteristics, the intentionality (*yi*) of [their] Heaven-given senses toward the phenomenon also is the same.... Articulated form and patterns of appearance are differentiated by the eye; the striking *sounds* in the notes and tunings are differentiated by the ear; the striking *flavors* among the sweet, bitter, salty, bland, hot, and sour are differentiated by the mouth; the striking odors among the fragrant, foul, grassy, florid, rotten, and putrid smells are differentiated by the nose. The painful, itchy, cool, hot, smooth, rough, light and heavy are differentiated by the body. The desires arising from joy, anger, sorrow, delight, love and hate are differentiated by the heart. In the heart there is a verifying faculty. This verifying faculty relies on the ears and then knowledge of sound is possible; it relies on the eyes for knowledge of forms to be possible. However, this verifying faculty must wait for the Heaven-given senses to register the category, and only then can it work.[17]

---

persistent topic of inquiry, with the range of interpretations extending from a strongly religious reverence for Heaven to the completely neutral "nature." John H. Berthrong's "Religion in the *Xunzi*: What Does *Tian* 天 Have to Do with It?" (*Dao Companion to the Philosophy of Xunzi*, pp. 323-51) provides a good assessment. However, for the main competing views, see Edward J. Machle, *Nature and Heaven in the Xunzi: A Study of the* Tian Lun (Albany: SUNY Press, 1993) for a strongly religious perspective; Paul Rakita Goldin's *Rituals of the Way: The Philosophy of Xunzi* (Chicago: Open Court Publishing Company, 1999), which draws a comparison with Deist models; and Kurtis Hagen, *The Philosophy of Xunzi: A Reconstruction* (Chicago: Open Court Publishing Company, 2007) for a largely human-centered analysis. Janghee Lee's excellent study, *Xunzi and Early Chinese Naturalism* (Albany: SUNY Press, 2005), presents Xun Kuang's philosophical formulations—and centrally, his approach to *Tian*—as a sophisticated response to the emerging forms of naturalism in the late Warring States.

[17] "On the Rectification of Names" 《正名》 in Wang Xianqian 王先謙, *Xun Zi jijie* 荀子集解 (Beijing: Zhonghua shuju, 1997), pp. 415-18. Cf. Eric L. Hutton, *Xunzi: The Complete Text* (Princeton: Princeton University Press, 2014), p. 238. Xun Kuang's philosophy of language has attracted significant scholarly attention. See, for example, Thomas D. Carroll, "Wittgenstein and the *Xunzi* on the Clarification of Language, (*Dao* 17 [2018]:527-45) and Chris Fraser, "Language and Logic in the *Xunzi*" (*Dao Companion to the Philosophy of Xunzi*, pp. 291-321.)

Xun Kuang argued that in this world of embodied, intentionally informed knowledge, we should accept the human limits of our capacity for knowledge:

> To not act yet complete, to not seek yet attain: this we call the assignment (職) of Heaven. [Heaven's assignment] being like this, although deep, one does not give thought to it; although vast, one does not put one's abilities to it; although rarified, one does not examine it: this we call not competing with Heaven in its assignment. The constellations follow their circuit, the sun and moon succeed one another in their brightness, the four seasons replace one another, Yin and Yang transform, wind and rain spread broadly, and each of the myriad phenomena attain [Heaven's] harmonizing in its birth and [Heaven's] nourishing in its completion. Not seeing the event but seeing its result: we call this the spirit[-driven]. We all know that by which [Heaven] completes [its transformations], yet none knows its being without form: we call this Heaven's merit.[18] Only the sage acts not to seek to know Heaven.[19]

To build his arguments, Xun Kuang uses the basic technical vocabulary shared among many thinkers of the late Warring States. Humans have a "nature" (性 *xing*) that derives from the active material (氣 *qi*) of which they are made, which is allotted by Heaven (天, *tian*). People's "natures" ground their responses to the world.[20] These responses are the *qing* 情, which, for humans, are largely understood as "dispositions" or "feelings."[21] Humans have a basic set of commitments

---

[18] There is a textual issue here. The standard modern edition of *Xun Zi* has 夫是之謂天 "we call this Heaven" rather than 天功 "Heaven's merit." Like Wang Niansun, I prefer the latter.

[19] "Discourse on Heaven" 《天論》 in Wang, *Xun Zi jijie*, pp. 308-09. Cf. Hutton, *Xunzi*, p.175-76.

[20] Dan Robins offers an overview of the usage of the term in "The Warring States Concept of *Xing*," *Dao* 10 (2011):31-51.

[21] Chad Hansen provocatively translates *qing* as "reality feedback," which works very well for the "Great Preface," discussed below. See Chad Hansen, *A Daoist Theory of Chinese Thought: A Philosophical Interpretation* (Oxford: Oxford University Press, 1992), pp. 276-77. More recently, Curie Virag defined the range of her inquiry in *The Emotions in Early Chinese Philosophy* to account for the semantic breadth of the term *qing*.

about the world ("resolve," 志 *zhi*) that, upon encountering particular objects and events, creates specific "intentions" (意 *yi*). Another early text describes the interactions among these components that were part of a shared understanding during the period:

> For any person, although he has a Nature, his mind is without a settled resolve. It waits upon objects before it becomes active. It awaits joy before it acts and awaits repeated practice before it is settled. The *qi* of joy, anger, sorrow and sadness are in one's Nature. When they are seen outside, it is because objects have elicited them. The Nature comes from one's fated allotment, and allotment descends from Heaven.[22]

Xun Kuang and other late Warring States Confucian thinkers describe a world in which we humans, made of material bestowed by a transcendent Heaven, know the phenomenal realm through our senses and thus know the world *as humans*, following distinctive human interests. Human nature is fundamentally reactive: it has commitments that compel its responsive engagement with the world as it is known through the human faculties of knowledge. Moreover, human nature itself, being inward, is only knowable through the patterns of those responses.[23]

> Instead, I will deploy the term "emotion" as my general category of reference. By this I refer to an entire spectrum of interrelated phenomena, from involuntary, physiological responses to external stimuli, to basic human dispositions and inclinations, to forms of cognition and perceptions of meaning, and to judgments and evaluations.

See her discussion in Curie Virag, *The Emotions in Early Chinese Philosophy* (Oxford: Oxford University Press, 2017), pp. 6-7. Most scholars writing in English who focus on Xun Kuang translate *qing* using terms like "dispositions," "conditions," or "human inclinations" and consistently argue that the term "emotions" is too narrow to reflect the breadth of meaning of *qing*. Eric L. Hutton discusses *qing* in Xun Kuang in "Xunzi on Moral Psychology" (*Dao Companion to the Philosophy of Xunzi*, pp. 201-27).

[22] "The Nature emerges from Allotment," section 1. Cf. the translation by Kenneth W. Holloway in *The Quest for Ecstatic Morality in Early China* (Oxford: Oxford University Press, 2013), p. 114. Holloway has an extensive discussion of this text.

[23] "The Nature emerges from Allotment"—a text that probably was written around 300 BCE (before Xun Kuang's active period)—was recovered from a tomb in 1993, and there is no

The neurobiological framework presented in this book has close correspondences with this model of knowing in the early Chinese tradition. And in premodern China poetry, which served as a crucial medium for the exploration of embodied, phenomenal meaning and identity, offers us much on which to reflect in our own construction of meaning within human, embodied terms.

The Poetry of Encounter

Poetry in premodern China is, at core, a poetry of encounter: writers, engaging objects and events, used poetry to structure their responses as moments of articulated meaning. This poetic tradition grew from a model of composition built on the early Chinese matrix of beliefs about human nature, its inwardness, and its responsiveness to events through the stirring of emotions.[24] The late Han Dynasty (206 BCE–220 CE) writers who began to use popular song forms to articulate their engagement with events took their model from the "Great Preface" to the *Canon of Poetry*, the revered repository of ancient verse deemed to be morally and aesthetically normative. The "Great Preface," probably written in the early Han Dynasty, begins:

> Poetry is where the resolve goes. While in the heart, it is the resolve; manifested in words, it is a poem. Emotion moves within and takes

---

evidence of its wider circulation. However, the "Record of Music" in the *Record of the Rites*, a Confucian canonical text, shares the same view of Nature being still until moved by objects and events. When the Nature is moved, it produces feelings:

> Being quiescent at birth is part of the human Heaven-granted Nature. Being moved by phenomenal objects and being stirred derives from the desires of the Nature. Objects arrive and the faculty of knowledge knows this, and only then do likes and dislikes form.

For the relation between "The Nature emerges from Allotment," Xun Kuang's thought, and the "Record of Music," see Franklin Perkins, "Music and Affect: The Influence of the *Xing Zi Ming Chu* on the *Xunzi* and *Yueji*," *Dao* 16 (2017):325-40.

[24] Goldin in *Rituals of the Way* draws an explicit connection between *Xun Zi*, the "Record of Music," and the "Great Preface" (pp. 77-81). Michael Nylan argues more generally that the Han dynasty elite' understanding of human nature was grounded in Xun Kuang's model and that his views remained central cultural commitments through the Tang (Michael Nylan, "Xunzi: An Early Reception History, Han Through Tang," *Dao Companion to the Philosophy of Xunzi*, pp. 395-433.)

shape in words. Words are not enough, and so one sighs it. Sighing it is not enough, and so one draws it out in song. Drawing it out in song is not enough, and so all unawares one's hands dance it and one's feet tap it out. Emotion is manifested in the voice.[25]

The preface answers the question "Why poetry?" Its argument is that the resolve one holds in one's heart is inward, as are one's "emotional" responses to encountered events. "Emotion" in the preface is *qing*, but the term is broader than "emotion" in English and encompasses a range of internal responses that include both emotions and intentions and accords well with the neuroscientific account presented in this book. These internal responses "naturally" require external manifestation.[26] For intense responses, mere words did not provide an adequate external form, and so one drew on the greater representational power of poetry—which in ancient China included both song and dance—to express one's response. Thus, in the schema provided by the "Great Preface," poetry is bound to its occasion as the external manifestation of the writer's intense internal response to the encountered event, a response complicated enough to require not merely simple words but more complexly structured expression.

Little poetry survives from the three hundred years between the writing of the "Great Preface" and the appropriation of popular song forms by the late Han literati who sought to shape a renewed poetic tradition. It was by no means inevitable that these writers would draw on the Preface's understanding of poetic

---

[25] See the discussions of this preface in Steven Van Zoeren, *Poetry and Personality: Reading, Exegesis, and Hermeneutics in Traditional China* (Stanford: Stanford University Press, 1991), pp. 80-115; Stephen Owen, *Readings in Chinese Literary Thought* (Cambridge: Harvard Council on East Asian Studies, 1992), pp. 37-56; and Bruce Rusk, *Critics and Commentators: The* Book of Poems *as Classic and Literature* (Cambridge: Harvard Asia Center Press, 2012), pp. 27-28.

[26] Xun Kuang, in his "Discourse on Music," uses the similarity between "music" (*ngaewk* 樂) and "to delight in" (*ngaewh* 樂) to explain:

> Now music is joy, something human feelings cannot avoid. Thus, people cannot be without joy, and joy must manifest itself in sound and take form in movement. And in the Human Way, the transformations of the operations of Nature are complete in sound and movement. Therefore people cannot but know joy, and joy cannot but take external form.

composition, and yet they did and, quite remarkably, found that it suited their purposes in ways that proved viable for the next thousand years.[27] In explaining how a poetics of external manifestation of internal responses came to meet important cultural needs, one additional factor based on early ontological assumptions may have been at work. While religious beliefs allowed that people, while alive, might transform and ascend to a Heaven as immortals (or descend to the Yellow Springs after death) and while most writers shared the belief in two souls—one that returned to the earth and another that needed the propitiation of sacrifice—the sense of the continuity of personal existence beyond the grave was weak and seemingly offered no consolation. Authors, instead, began to see that a form of immortality was possible through writing. Cao Pi (187-226), who was a key figure in the group of writers who shaped the beginning of the new poetry, as well as the future emperor of the Wei Dynasty, ended his "Discourse on Writing" with reflections on writing and mortality:

> Literary works are the greatest accomplishment in the workings of a state, a splendor that never decays. Glory and pleasure go no further than this mortal body. To extend both of these to all time, nothing can compare with the unending permanence of a work of literature. It was for this reason that writers of ancient times gave their lives to the ink and brush and revealed what they thought in their writings. Without recourse to a good historian or dependence on a powerful patron, their reputations have been passed down to posterity on their own force.... We can see from this that the ancients cared nothing for those great jade disks that were marks of wealth, but instead treasured the moment, fearful lest time pass them by. Yet people tend not to exert themselves in this way. In poverty and low position, they fear hunger and cold; when rich and honored, they let themselves drift in the distractions of pleasure. They occupy themselves with immediate demands and neglect an accomplishment that will last a thousand

---

[27] In a major epistemic shift marked by the rise of Neo-Confucianism in the Southern Song Dynasty, *qing* as a category of human experience was largely reduced simply to the passions and lost any significant role in elucidating the meaning of the human engagement with the world. The poetic tradition was profoundly affected—and weakened—in this cultural transformation.

years. The days and months pass overhead; here below the face and body waste away. We will pass suddenly into change with all the things of this world—and this causes great pain to a man with high aspirations.[28]

Among all the literary genres, poetry, expressing a response to encountered events, most directly preserved a record of one's distinctive individuality for later times. Thus Cao Zhi (192-232), the most talented poet of the group, wrote poems that imitated the generic song styles that they adopted but also turned poetry into a more immediately expressive medium, as in this farewell poem.

送應氏二首其一　　*Seeing off Mr. Ying*[29]

步登北邙坂，	On foot I ascend the slope at North Mang:[30]
遙望洛陽山。	Afar I gazed at the hills of Loyang.
洛陽何寂寞，	Loyang: how desolate!
宮室盡燒焚。	The buildings all utterly burnt.
垣牆皆頓擗，	The walls entirely cast down;
荊棘上參天。	Brambles rise to touch the sky.
不見舊耆老，	I do not see the old ones of former times.
但睹新少年。	I look at the newly [arrived] youths.
側足無行徑，	Turning my feet sideways, there are no paths to travel.
荒疇不復田。	Overgrown fields are not cultivated again.
遊子久不歸，	A wandering one long has not returned:
不識陌與阡。	[You] do not recognize the paths and lanes.
中野何蕭條，	Amidst the wilds, how forlorn:

---

[28] Xiao Tong, *Wenxuan* (Taipei: Wenyi, 1974) 52.733-34. The translation is from Stephen Owen, *An Anthology of Chinese Literature: Beginnings to 1911* (New York: W. W. Norton and Company, 1996), pp. 360-61. For a more extensive analysis, see Owen, *Readings in Chinese Literary Thought*, pp. 57-72. Xiaofei Tian discusses the context of the plague within which Cao Pi wrote the "Discourse on Writing" in *The Halberd at Red Cliff: Jian'an and the Three Kingdoms* (Cambridge: Harvard University Asia Center Press, 2018), pp. 13-30.

[29] Cao Zhi, *Cao Zhi ji jiaozhu*, edited by Zhao Youwen (Beijing: Renmin wenxue, 1998) 1.3. The translation is from Michael A. Fuller, *An Introduction to Chinese Poetry: from the* Canon of Poetry *to the Lyrics of the Song Dynasty* (Cambridge: Harvard University Asia Center Press, 2017), p. 117.

[30] North Mang was a hill to the northeast of Loyang, the capital of the Eastern Han, where many of the nobility had their grave sites.

千里無人煙。	For a thousand *li*, no smoke of human habitation.
念我平生親，	Recalling one with whom I have been close all my life,
氣結不能言。	My breath constricts, and I cannot speak.

The task of the poem is to articulate the meaning of Mr. Ying's departure. This "meaning" is a nexus of intersecting concerns: the capital has been ravaged, and much has changed; those he knew of old are mostly gone, replaced by strangers; and now Mr. Ying sets out across a barren landscape. The *qing*—no simple single object—is given voice through a set of affectively freighted observations. Still, this is a poem rather than reportage. The couplets rhyme; the imagery is selected and structured; and the ending relies on a form of closure common in the poetry of the period: "I am so moved that my still unexpressed emotions leave me in silence." Technical aspects of composition continued to evolve over the following centuries as poets expanded both the range of topics and the rhetorical tools for structuring meaning, but the core of the tradition remained poems reflecting on events.

    I suggest that if we want to think about the nature of meaning and identity as they are recast in the neurobiological framework, we can attend to the poets of the classical Chinese tradition, who worked within an understanding of the human condition that mirrors the neuroscientific model and developed a rich, compelling vision of the world of meaning that surrounds us that is worthy of our reflection. In the framework developed in this book, both perception and memory are fundamentally committed to capturing the significant patterns of the world and the self, but these are patterns of *human* significance inflected by creaturely affective commitments that go far beyond simple bodily requirements. I would argue that the poetry of the classical Chinese tradition in effect applies this neuroscientific model in its interpretive engagement with the world of experience. "Weary Night," a poem by Du Fu (713-770), the greatest poet in the tradition, for example, brings this book's phenomenally constrained, embodied framework to life in a quiet yet powerful way.

倦夜	*Weary Night*[31]
竹涼侵臥內	Bamboo coolness invades the bedroom;
野月滿庭隅	The outland moon fills a corner of the courtyard.
重露成涓滴	The heavy dew forms water drops that fall.
稀星乍有無	The sparse stars flicker, now there, now gone.
暗飛螢自照	Fireflies gleam in their own light as they fly in the dark.
水宿鳥相呼	The birds stopping for the night on the water call out to one another.
萬事干戈裏	Ten thousand affairs all hemmed in with weapons:
空悲清夜徂	In vain I sorrow that the pure night passes.

Du Fu at the time was in Chengdu, Sichuan, where he had moved his family to support and protect them during the chaotic upheaval of the disastrous An Lushan Rebellion.[32] This is a poem of "indefinite occasion" reflecting on—and structuring the meaning of—the experience of a sleepless night. The poem is filled with patterns that Du Fu discovers in the scene before him, patterns through which he situates himself within the broader—though deeply personal—significance of a moment that is complexly, inextricably bound to larger temporal processes. In his reflections on the quiet summer night, perception, personal emotion, and reflective assessment blend together. It is late summer, and the coolness of the air drifting through the bamboo into his bedroom is a welcome relief, even if Du Fu introduces the discordant note of "invading." The sharp shadow of the bright moon creates a scene in his small garden half in darkness, half in moonlight. Still, even as he focuses on this scene, he imaginatively projects beyond it, since this is an "outland" moon coming to his chambers. Spaces are defined if not actually seen. The two parallel inner couplets of the poem note patterns of seasonal transformation encompassing this

---

[31] Fuller, *An Introduction to Chinese Poetry*, pp. 252-53. Cf. Stephen Owen, trans. and ed., *The Poetry of Du Fu* (Boston: De Gruyter, 2016), 4.12.

[32] For an introduction to Du Fu's life and poetry, see my *An Introduction to Chinese Poetry*, pp. 220-68.

place at this time.[33] The dew appears because of the new nighttime coolness, and Du Fu sees the glinting because of the brightness of the moon that night. The pattern of the flickering of dewdrops that disappear as they fall, part of the near scene, are mirrored by the stars, small points of light far above. They are sparse because the moon is bright, and they flicker because of the dense, humid atmosphere. Du Fu then introduces a new source of small, flickering light: the fireflies wandering in the shaded part of his garden are, like Du Fu, restless and reveal themselves. Since this is late summer, the fireflies will die soon, lending a quiet urgency to their searching. Paired with the visual image of the fireflies, the migratory birds on the nearby river, also awake in the night, are heard rather than seen, but they are not alone as they rest from their seasonal travel. Du Fu sees their intermittent calls as part of the same larger pattern as the flashing of the fireflies, and he implicitly recognizes his own composing of the poem responding to transformation as parallel to the visual and aural acts of communication of the birds and fireflies.

The final couplet shifts these patterns of restlessness and transformation to the "seasons of man" through which Du Fu has paid particular attention to elements of the scene that resonate with his abiding concern. The empire is still at war, while time passes and he accomplishes little. The final line—"In vain I sorrow that the pure night passes"—is Du Fu's response to all that has proceeded it in the poem and expresses a complex mood that shows the power of poetry. The night indeed has presented a "pure" scene, which he acknowledges, and while he regrets that it is coming to an end, he knows that both his appreciation and his regret are in vain. They seemingly accomplished nothing to change his situation. And yet something has changed: through the "work" of the poem, the introduction, structuring, and shifting in the patterns of images, the human pattern—the continuing warfare and Du Fu's insignificant role in it—become part of a yet larger world in transformation, filled with purity, regret, and restlessness.

---

[33] The poem is in "regulated verse" format, so that the words in the two lines of each of the inner couplets have some form of categorical correspondence. Du Fu in the poem exploits the expectation of parallel structure to compel the reader to seek out the logic of correspondence in a way that is part of the pleasure, interest, and success of the poem.

"Weary Night," as a poem, has a complex semantic structure and cannot usefully be simplified to a short set of propositions.[34] As in the neuroscientific framework that I have proposed, meaning in the poem radiates from the interaction between words as centers of semantic structures. The images evoke complex memories that infuse affective elements into humanly relevant predictive knowledge of the patterns of experience. When Du Fu structures these images into the matrix of the poem, he brings particular aspects of the ramifying memory networks to the fore. The result is a rich web of meanings connecting the author and the reader to the details of the encountered world that emerge in experience. The poetic tradition taught generations of Chinese literati how to read meaning in experience, how to see themselves in a continuity with the patterns of the phenomenal realm, which they could then reimagine in their own lives. The neuroscience suggests that we as individuals, from the moment of birth, do indeed participate in an analogous matrix of meaningful structures in both a social world and the humanly mediated phenomenal realm of which the human world is a small part. The challenge as we engage the neuroscientific models is not merely to see the matrix of meaning but to realize that, as in the Chinese use of their poetic tradition, we can imagine new structures within the vast possibilities of the patterns of the world and arrange them in thought and action in ways that are compelling to us.

The neuroscientific framework for perception, emotion, and memory presented in this book do not and cannot explain Du Fu's "Weary Night." It should not be expected to: the distance between the experiential levels is simply too great; too many layers of emergent organization based on the complexities of individual experience and the historically shaped internalized (or resisted) structures of the world that form the grist of experience intervene between the neuronal structures and dynamics and such creative works as poems. But the neuroscience does offer

---

[34] Falke's observation effectively describes the experience of reading the poem:
> The insight-rooted knowledge that literature promotes enables a specific privileged relation to the world rather than conveying a collection of facts about it. It is thus knowledge we inhabit more than acquire.

Falke, *The Phenomenology of Love and Reading*, p. 53.

a powerful and flexible substructure for experience, meaning, and selfhood. It will not be relevant in all forms of inquiry, but when we probe deeply into the logic of human experience, it can help us think through fundamental structures. Within the confines of phenomenal experience, we are biological beings. If we want to imagine the possibilities for who we can become, we need to be part of the larger discourse of the human that necessarily includes the neurobiological dimension of our existence. I have written this book in hopes of providing a conceptual vocabulary to assist in nurturing this much-needed conversation.

Permissions

Chapter One

Figure 1.1: The Neuron
  Brett Szymik. "A Nervous Journey." http://askabiologist.asu.edu/neuron-anatomy. © Arizona Board of Regents / ASU Ask A Biologist. Published under Creative Commons BY-SA 3.0.

Figure 1.3a: "Winner takes all" Neural Network
  Rumelhart, David E., James L. McClelland, and PDP Research Group, *Parallel Distributed Processing, Volume 1*, Figure 5, page 172, © 1986 Massachusetts Institute of Technology, by permission of The MIT Press.

Figure 1.3b: "Winner takes all" Neural Network Trial Results
  Rumelhart, David E., James L. McClelland, and PDP Research Group, *Parallel Distributed Processing, Volume 1*, Figure 6, page 173, © 1986 Massachusetts Institute of Technology, by permission of The MIT Press.

Figure 1.4: Model of Neural Network connecting the Lateral Geniculate
  Steven J. Olson and Stephen Grossberg, "A neural network model for the development of simple and complex receptive fields within cortical maps of orientation and ocular dominance," *Neural Networks* 11 (1998):189-208. © 1996 Boston University, published under CC BY.

Figure 1.5: Kohonen Self-organizing Map Network
  © Jochen Fröhlich, Neural Networks with Java, https://www.nnwj.de/kohonen-feature-map.html. Permission requested Nov. 16, 2022.

Figure 1.6: "Mexican Hat" Weight function for Lateral Connections in SOM
  A. Mwegerano, P. Kytösaho and A. Tuominen, "Applying Self-Organizing Maps Method to Analyze the Corrective Action's Quality Provided to Customers with Mobile Terminals," *iBusiness*, Vol. 4 No. 2, 2012, Figure 7 p. 111. doi: 10.4236/ib.2012.42013. © SCIRP, 2012, Published under Creative Commons BY.

Figure 1.7a: Architecture of the LISSOM model
  Philips RT and Chakravarthy VS, "A Global Orientation Map in the Primary Visual Cortex (V1): Could a Self-Organizing Model Reveal Its Hidden Bias?" *Frontiers in Neural Circuits* 10.109 (January 2017), Figure 1, p. 4. © The authors, published under Creative Commons BY, 2017

Figure 1.7b: LISSOM Topological Feature Map Model for Line Orientation in Hypercolumns in V1
Philips and Chakravarthy, "A Global Orientation Map in the Primary Visual Cortex (V1)" *Frontiers in Neural Circuits*, Figure 11, p. 11. © The authors, published under Creative Commons BY, 2017

Figure 1.8: Model of Recurrent Connectivity in the Visual Cortices
Micah Richert, Dimitry Fisher, Filip Piekniewski, Eugene M. Izhikevich, and Todd L. Hylton, "Fundamental principles of cortical computation: unsupervised learning with prediction, compression and feedback" arXiv:1608.06277 [cs.CV] (August 2016), Figure 1, p. 6. © The authors, published under Creative Commons BY, 2016.

Figure 1.9: Attractor Basin
O'Reilly, R. C., Munakata, Y., Frank, M. J., Hazy, T. E., and Contributors (2020). *Computational Cognitive Neuroscience*. Wiki Book, 4th Edition. URL: https://CompCogNeuro.org, Figure 3.15, p. 55. © The authors, published under Creative Commons BY, 2020.

Figure 1.10: Figure-ground Discrimination
Richard L. Gregory, "The Medawar Lecture 2001, Knowledge for vision: vision for knowledge," *Philosophical Transactions of the Royal Society London B Biological Science* 360 .1458 (2005), figure 9, p. 1238. © Royal Society of London, 2005, published with permission of the Royal Society of London.

Figure 1.11: Attractor Basin Landscape
Chris Eliasmith, "Attractor Networks" in *Scholarpedia* 2(10):1380. http://www.scholarpedia.org/article/Attractor_network. © The author, published under Creative Commons Attribution-NonCommercial-ShareAlike 3.0 Unported License.

Figure 1.12: The Adaptive Resonance Model
© Chris McBain, https://chrismcbain.files.wordpress.com/2015/06/art4.png, accessed 11/18/2022. Permission requested Nov. 18, 2022.

Chapter Two

Figure 2.1: The Human Brain
Sami Azam, *Detection of binaural processing in the human brain*, Doctoral Thesis, Charles Darwin University, 2016. Page 30, Figure. 2.5. https://doi.org/10.25913/5ea243abbbd62.

Figure 2.2: Medial View of the Brain
https://www.123rf.com/profile_hfsimaging © 123rf.com, published with permission of 123rf.com.

Figure 2.3: The Lateral View of the Neocortex
Bernard J. Baars and Nicole M. Gage, *Cognition, Brain, and Consciousness: Introduction to Cognitive Neuroscience* second edition (New York: Academic Press, 2010), p. 145. © Academic Press, published with permission of Academic Press.

Figure 2.4: Names of Directions for the Brain
Bernard J. Baars and Nicole M. Gage, *Cognition, Brain, and Consciousness: Introduction to Cognitive Neuroscience* second edition (New York: Academic Press, 2010), p. 145. © Academic Press, published with permission of Academic Press.

Figure 2.5: The 6-Layered Neocortex
Figure III-10-5, The 6-Layered Neocortex (p. 725) in *USMLE Step 1: Anatomy – Lecture Notes*, edited by James White and David Seiden (Kaplan Medical Books, 2018). © Simon & Schuster, 2018. Permission request letter sent Nov. 20, 2022.

Figure 2.7a: Sensory Homunculus
Wilder Penfield and T. Rasmussen, *The Cerebral Cortex of Man: A Clinical Study of Localization of Function* (New York: Macmillan, 1957), Figure 17, p.44. Current copyright holder not located.

Figure 2.7b: Motor Homunculus
Wilder Penfield and T. Rasmussen, *The Cerebral Cortex of Man: A Clinical Study of Localization of Function* (New York: Macmillan, 1957), Figure 22, p.57. Current copyright holder not located.

Figure 2.8: White Matter Tracts
Tom Barrick, Chris Clark, SGHMS/ Science Photo Library / Getty Images Plus, https://www.sciencephoto.com/media/307225/view. © Science Photo Library, published with permission.

Figure 2.9: Major Networks in the Brain
Olaf Sporns, "Network attributes for segregation and integration in the human brain," *Current Opinion in Neurobiology* 23 (2013), p. 167. The image is Figure 3, p. 166. © Elsevier Ltd. 2013, published with permission of Elsevier Ltd.

Chapter Three

Figure 3.1: The Human Eye
Figure 2.6.1 of "The Eye" in *University Physics* (OpenStax) (2020, November 5), https://phys.libretexts.org/@go/page/4496, © OpenStax CNX, shared under a CC BY 4.0 license.

Figure 3.2: The Neuronal Structure of the Retina
The figure in Brittany J. Carr and William K. Stell, "The Science Behind Myopia", in Helga Kolb, Ralph Nelson, Eduardo Fernandez, and Bryan Jones, eds., *Webvision: The Organization of the Retina and Visual System* [Internet]. (Salt Lake City (UT): University of Utah Health Sciences Center; 1995- present). https://webvision.med.utah.edu/book/part-xvii-refractive-errors/the-science-behind-myopia-by-brittany-j-carr-and-william-k-stell/. is a modified version of Figure 2 in Helga Kolb, "Simple Anatomy of the Retina", in Helga Kolb, Ralph Nelson, Eduardo Fernandez, and Bryan Jones, eds., *Webvision:* https://webvision.med.utah.edu/imageswv/schem.jpeg, accessed 11/24/2022. © Helga Kolb, licensed through Creative Commons BY.

Figure 3.3: The Fovea
"Anatomy of the fovea - English labels" at AnatomyTOOL.org, © Cenveo, licensed through Creative Commons Attribution.

Figure 3.4: Using eye-tracking to record saccades
Alfred L. Yarbus, *Eye Movement and Vision*, trans. by Basil Haigh (New York: Plenum Press, 1967), Fig. 114, p 179. © Plenum Press, 1967, published with permission of Plenum Press.

Figure 3.5: Photoreceptor Response Sensitivity
Vectorized version of the GFDL image Cone-response.png uploaded by User:Maxim Razin based on work by User:DrBob and User:Zeimusu., Wikimedia Commons, https://commons.wikimedia.org/wiki/File:Cone-response-n.svg. © Wikimedia Commons, licensed under CC BY-SA 3.0.

Figure 3.6: The Center-Surround Model
Baars and Gage, *Cognition, Brain, and Consciousness*, p. 162, Figure 6.4. © Academic Press, published with permission of Academic Press.

Figure 3.7: Image of Filtering to enhance edges in Ganglion Cells
Baars and Gage, *Cognition, Brain, and Consciousness*, p. 162, Figure 6.5. © Academic Press, published with permission of Academic Press.

Figure 3.8: Binocular Visual Pathways
E. Herrera and C.A. Mason, "3.23 - The Evolution of Crossed and Uncrossed Retinal Pathways in Mammals," in *Evolution of Nervous Systems* Volume 3 (2007), Figure 2. © Elsevier Ltd. 2007, published with permission.

Figure 3.9: Orientation Selectivity in V1
> Baars and Gage, *Cognition, Brain, and Consciousness*, p. 164, Figure 6.8. © Academic Press, published with permission of Academic Press.

Figure 3.10: Ocular Dominance Columns
> McGill: *The Brain from Top to Bottom*, Figure of hypercolumns, Copyleft https://thebrain.mcgill.ca/flash/a/a_02/a_02_cl/a_02_cl_vis/a_02_cl_vis.html.

Figure 3.11: Increasing Complexity of Feature-Map Structural Units
> Baars and Gage, *Cognition, Brain, and Consciousness*, p. 166, Figure 6.10. © Academic Press, published with permission of Academic Press.

Figure 3.12: The Visual Cortices
> Nikos K. Logothetis, "Vision: A Window into Consciousness," *Scientific American* 281.5 (Nov. 1999), p. 72. © Terese Winslow LLC, published with permission of the illustrator.

Figure 3.13: Reciprocal Connectivity to the Visual Cortex
> Charles D. Gilbert and Wu Li, "Top-down Influences on Visual Processing" *Nature Reviews Neuroscience* 14 (May 2013), p. 351. © Springer Nature Ltd. 2013, published with permission.

Figure 3.14: Neuronal Control Networks and Hubs
> Caterina Gratton, Haoxin Sun, and Steven E. Petersen, "Control Networks and Hubs" *Psychophysiology* 55.3 (March 2018), DOI 10-1111/psyp.13032, p. 7, Figure 2. © John Wiley and Sons, Inc. 2018, published with permission.

Figure 3.15: The Dorsal and Ventral Streams
> © Selket - I (Selket) made this from File:Gray728.svg, https://commons.wikimedia.org/w/index.php?curid=1679336, © Wikimedia Commons, licensed under CC BY-SA 3.0.

Chapter Four

Figure 4.1: Affective Hierarchies
> "Cross-Species Affective Neuroscience Decoding of the Primal Affective Experiences of Humans and Related Animals," *PLoS One* 6.9 (September 2011), p. 6. © PLoS, licensed under Creative Commons Attribution 4.0 International (CC BY).

Figure 4.2: The Human Parenting Network
> Ruth Feldman, Katharina Braun and Frances A. Champagne, "The neural mechanisms and consequences of paternal caregiving," *Nature Reviews:*

Neuroscience 20 (April 2019), p. 210. © Springer Nature Ltd. 2019, published with permission.

Figure 4.3ab: The Visual Perceptual Hierarchy
   Ryan Smith and Richard D. Lane, "The neural basis of one's own conscious and unconscious emotional states," *Neuroscience and Biobehavioral Reviews* 57 (2015), Figure 2, p. 8. © Elsevier Ltd. 2015, published with permission.

Figure 4.4: Affective Networks of the Brain
   Ryan Smith and Richard D. Lane, "The neural basis of one's own conscious and unconscious emotional states," *Neuroscience and Biobehavioral Reviews* 57 (2015), p. 17. © Elsevier Ltd. 2015, published with permission.

Chapter Five

Figure 5.1: The Fetal Forming of the Layers of the Cortex
   Joan Stiles, Timothy T. Brown, Frank Haist, and Terry L. Jernigan, "Brain and Cognitive Development," in Richard M. Lerner, ed., *Handbook of Child Psychology and Developmental Science, 7th Edition*, (John Wiley & Sons, Inc., 2015), Figure 2.4, p. 12. © John Wiley and Sons, Inc. 2018, published with permission.

Figure 5.2: Developmental Timeline
   John H. Gilmore, Rebecca C. Knickmeyer and Wei Gao, "Imaging structural and functional brain development in early childhood," *Nature Reviews Neuroscience* 19 (March 2018), p. 127. © Springer Nature Ltd. 2018, published with permission.

Figure 5.3: Neuronal Developmental Timeline
   Baars and Gage, *Cognition, Brain, and Consciousness*, Figure 15.15, p. 479. © Academic Press, published with permission of Academic Press.

Chapter Six

Figure 6.1: An Artificially Generated Dendrogram
   Based on James L. McClelland, Bruce L. McNaughton, and Randall C. O'Reilly, "Why There Are Complementary Learning Systems in the Hippocampus and Neocortex: Insights from the Successes and Failures of Connectionist Models of Learning and Memory," *Psychological Review* 102.3 (1995): 431, Figure 7.

Figure 6.2: Distribution of Memory in the Brain
   Joaquin Fuster, "Upper Processing Stages of the Perception-Action Cycle,"

Trends in Cognitive Science 8.4 (April 2004), p. 144. © Elsevier Ltd. 2004, published with permission.

Figure 6.3: The Medial Temporal Lobe Structure
F.D. Raslau, I.T. Mark, A.P. Klein, J.L. Ulmer, V. Mathews and L.P. Mark, "Memory Part 2: The Role of the Medial Temporal Lobe" *American Journal of Neuroradiology* May 2015, 36 (5) 846-849; DOI: https://doi.org/10.3174/ajnr.A4169, Figure 1, p. 846. © American Society of Neuroradiology 2015, published with permission.

Figure 6.4: The Two Pathways in the Hippocampus
Thomas Hainmueller and Marlene Bartos, "Dentate gyrus circuits for encoding, retrieval and discrimination of episodic memories," *Nature Reviews Neuroscience* 21.3 (March 2020), p. 154. © Springer Nature Ltd. 2020, published with permission.

Figure 6.5: The Mossy Fiber- CA3 Synapse
Nelson Rabola, Mario Carta, and Christophe Mulle, "Operation and plasticity of hippocampal CA3 circuits: implications for memory encoding," *Nature Reviews Neuroscience* 18.4 (April 2017), Figure 2, p. 211. © Springer Nature Ltd. 2017, published with permission.

Figure 6.6: The Connections to CA3 in the Hippocampus
Rabola et al., "Operation and plasticity of hippocampal CA3 circuits: implications for memory encoding," Figure 1, p. 210. © Springer Nature Ltd. 2017, published with permission.

Figure 6.7: The Sleep Cycle
Jens G. Klinzing, Niels Niethard and Jan Born, "Mechanisms of systems memory consolidation during sleep," *Nature Neuroscience* 22 (October 2019), p. 1603, Box 3 Figure. © Springer Nature Ltd. 2019, published with permission.

Figure 6.8: Memory Consolidation
Klinzing et al., "Mechanisms of systems memory consolidation during sleep," p. 1600, Figure 1. © Springer Nature Ltd. 2019, published with permission.

Figure 6.9: Neural Networks to Support Language Perception and Production
Gregory Hickok and David Poeppel, "Neural Basis of Speech Perception," in Gregory Hickok and Steven L. Small, eds., *Neurobiology of Language*, (London: Academic Press, 2015), p. 300. © Academic Press, 2015, published with permission.

Figure 6.10: Dependency Parsing
Alessandro Lopopolo, Antal van den Bosch, Karl-Magnus Petersson, and

Roel M. Willems, "Distinguishing Syntactic Operations in the Brain: Dependency and Phrase-Structure Parsing," *Neurobiology of Language* 2.1 (January 2021), p. 157, Figure 2. © 2021 Massachusetts Institute of Technology. Published under CC BY 4.0 license.

Figure 6.11: Phrase-Structure Parsing

Lopopolo, "Distinguishing Syntactic Operations in the Brain," p. 156, Figure 1. © 2021 Massachusetts Institute of Technology. Published under a CC BY 4.0 license.

Figure 6.12: Connectivity in the Language System

Angel D. Friederici, "The Neuroanatomical Pathway Model of Language: Syntactic and Semantic Networks," Hickok and Small, eds., *Neurobiology of Language*, p. 351. © Academic Press, 2015, published with permission.

Bibliography

BOOKS

Asma, Stephen T. and Rami Gabriel. *The Emotional Mind: The Affective Roots of Culture and Cognition.* Cambridge: Harvard University Press, 2019.

Baars, Bernard J. and Nicole M. Gage. *Cognition, Brain, and Consciousness: Introduction to Cognitive Neuroscience* second edition. Boston: Academic Press, 2010.

___. Fundamentals of Cognitive Neuroscience: A Beginner's Guide. Boston: Academic Press, 2013.

Barrett, Karen Caplovitz, Nathan A. Fox, George A. Morgan, Deborah J. Fidler, Lisa A. Daunhauer, eds. *Handbook of Self-Regulatory Processes in Development: New Directions and International Perspectives.* New York: Psychology Press, 2013.

Barrett, Lisa Feldman and James A. Russell, eds. *The Psychological Construction of Emotion.* New York: Guilford Press, 2015.

Bear, Mark, Barry W. Connors, Michael A. Paradiso. *Neuroscience: Exploring the Brain.* Philadelphia: Wolters-Kluwer, 2016.

Boddice, Rob. *The History of Emotions.* Manchester: Manchester University Press, 2018.

Cassidy, Jude and Phillip R. Sharver, eds. *Handbook of Attachment: Theory, Research, and Clinical Application*, 3rd edition. New York: The Guilford Press, 2016.

Clark, Andy. Surfing Uncertainty: Prediction, Action, and the Embodied Mind. Oxford: Oxford University Press, 2016.

Clark, Edwin and Kenneth Dewhurst. *An Illustrated History of Brain Function* 2nd edition. San Francisco: Norman, 1996.

Damasio, Antonio. Self Comes to Mind: Constructing the Conscious Brain. New York: Vintage, 2012.

Dilthey, Wilhelm. *Understanding the Human World, Volume II of Selected Works.* Edited by R.A. Makkreel and F. Rodi. Princeton: Princeton University Press, 2010.

Dilthey, Wilhelm. *The Formation of the Historical World in the Human Sciences, Volume III of Selected Works.* Edited by R.A. Makkreel and F. Rodi. Princeton: Princeton University Press, 2002.

Dowling, John E. *Creating Mind.* New York: Norton, 1998.

Du Fu. *The Poetry of Du Fu.* Translated and edited by Stephen Owen. 6 volumes. Boston: De Gruyter, 2016,

Falke, Cassandra. *The Phenomenology of Love and Reading.* New York: Bloomsbury Press, 2017.

Fuller, Michael A. *An Introduction to Chinese Poetry: from the* Canon of Poetry *to the Lyrics of the Song Dynasty.* Cambridge: Harvard University Asia Center Press, 2017.

Goldin, Paul Rakita. *Rituals of the Way: The Philosophy of Xunzi.* Chicago: Open Court Publishing Company, 1999.

Graziano, Michael S. A. Rethinking Consciousness: A Scientific Theory of Subjective Experience. New York: W. W. Norton & Company, 2019.

Hagen, Kurtis. *The Philosophy of Xunzi: A Reconstruction.* Chicago: Open Court Publishing Company, 2007.

Hansen, Chad. A Daoist Theory of Chinese Thought: A Philosophical Interpretation. Oxford: Oxford University Press, 1992.

Hickok, Greg and Steve Small, eds. *Neurobiology of Language.* Boston: Academic Press, 2015.

Holloway, Kenneth W. *The Quest for Ecstatic Morality in Early China.* Oxford: Oxford University Press, 2013.

Hubel, David. *Eye, Brain, and Vision.* New York: Scientific American Library, 1995.

Hutton, Eric L., ed. *Dao Companion to the Philosophy of Xunzi.* Dordrecht: Springer, 2016.

___, trans. *Xunzi: The Complete Text.* Princeton: Princeton University Press, 2014.

Johnson, Mark H. *Developmental Cognitive Neuroscience: An Introduction.* Oxford and Cambridge, MA: Blackwell, 1997.

Johnston, Adrian and Catherine Malabou. *Self and Emotional Life: Philosophy, Psychoanalysis, and Neuroscience.* New York: Columbia University Press, 2013.

Koch, Christof. The Feeling of Life Itself: Why Consciousness is Widespread but Can't Be Computed. Cambridge: MIT Press, 2019.

Lee, Janghee. *Xunzi and Early Chinese Naturalism*. Albany: SUNY Press, 2005.

Lerner, Richard M., ed. Handbook of Child Psychology and Developmental Science, 7th Edition. John Wiley & Sons, Inc., 2015.

Machle, Edward J. *Nature and Heaven in the Xunzi: A Study of the* Tian Lun. Albany: SUNY Press, 1993.

Marion, Jean-Luc. *The Erotic Phenomenon*. Translated by Stephen E. Lewis. Chicago: University of Chicago Press, 2007.

Mueller-Vollmer, Kurt, ed. The Hermeneutics Reader: Texts of the German Tradition from the Enlightenment to the Present. New York: Continuum Press, 1985.

Narvaez, Darcia, Jaak Panksepp, Allan N. Schore, and Tracy R. Gleason, eds. *Evolution, Early Experience and Human Development: From Research to Practice and Policy.* Oxford: Oxford University Press, 2013.

O'Reilly, R. C., Yuko Munakata, Michael J. Frank, Tom Hazy. *Computational Cognitive Neuroscience*. Wiki Book, 1st Edition. 2012. URL: http://ccnbook.colorado.edu.

Owen, Stephen. *An Anthology of Chinese Literature: Beginnings to 1911*. New York: W. W. Norton and Company, 1996.

———. *Readings in Chinese Literary Thought*. Cambridge: Harvard Council on East Asian Studies, 1992.

Panksepp, Jaak. Affective Neuroscience: The Foundations of Human and Animal Emotions. Oxford: Oxford University Press, 1998.

Penfield, Wilder and T. Rasmussen. The Cerebral Cortex of Man: A Clinical Study of Localization of Function. New York: Macmillan, 1957.

Plamper, Jan. *The History of Emotions*. Translated by Keith Tribe. Oxford: Oxford University Press, 2015.

Puji 普濟, ed. *Wudeng huiyuan* 五燈會元. Collated by Su Yuanlei. Beijing: Zhonghua, 1984.

Reddy, William. The Making of Romantic Love: Longing and Sexuality in Europe, South Asia & Japan, 900-1200 ce. Chicago: University of Chicago Press, 2012.

Rosenwein, Barbara. *Generations of Feeling: A History of Emotions, 600-1700*. Cambridge: Cambridge University Press, 2016.

Rosenwein, Barbara H. and Riccardo Cristiani. *What is the History of Emotions*. Cambridge: Polity Press, 2018.

Rumelhart, David E. and James L. McClelland, and PDP Research Group. *Parallel Distributed Processing: Explorations in the Microstructure of Cognition, Volume 1: Foundations*. Cambridge: MIT Press, 1986.

Rusk, Bruce. *Critics and Commentators: The Book of Poems as Classic and Literature*. Cambridge: Harvard Asia Center Press, 2012.

Stanghelline, Giovanni, Matthew Broome, Andrea Raballo, Anthony Vincent Fernandez, Paolo Fusar-Poli, and René Rosfort, eds. *The Oxford Handbook of Phenomenological Psychopathology*. Oxford: Oxford University Press, 2018. Online Publication: March 2018, DOI: 10.1093/oxfordhb/9780198803157.013.25.

Suzuki, Daisetsu Teitaro (D. T.) *Essays in Zen Buddhism*. New York: Grove Press, 1949.

Tian, Xiaofei. *The Halberd at Red Cliff: Jian'an and the Three Kingdoms*. Cambridge: Harvard University Asia Center Press, 2018.

Trappenberg, Thomas P. *Fundamentals of Computational Neuroscience* Second edition. Oxford: Oxford University Press, 2010.

Tsakiris, Manos and Helena De Preester, eds. *The Interoceptive Mind: From homeostasis to awareness*. Oxford: Oxford University Press, 2018.

Van Zoeren, Steven. Poetry and Personality: Reading, Exegesis, and Hermeneutics in Traditional China. Stanford: Stanford University Press, 1991.

Virág, Curie. *The Emotions in Early Chinese Philosophy*. Oxford: Oxford University Press, 2017.

Vygotsky, Lev. *Mind in Society: The Development of Higher Psychological Processes*. Edited by Michael Cole, Vera John-Steiner, Sylvia Scribner, and Ellen Souberman. Cambridge: Harvard University Press, 1978.

Wang, Xianqian 王先謙. *Xun Zi jijie* 荀子集解. Beijing: Zhonghua shuju, 1997.

Watts, Duncan J. *Six Degrees: The Science of a Connected Age*. New York: W.W. Norton and Company, 2003.

Xiao Tong 蕭統. *Wenxuan* 文選. Annotated by Li Shan 李善. Taipei: Wenyi, 1974.

Yarbus, Alfred L. *Eye Movement and Vision*. Translated by Basil Haigh. New York: Plenum Press, 1967.

Zahavi, Dan, ed. *The Oxford Handbook of Contemporary Phenomenology*. Oxford: Oxford University Press, 2012.

Ziporyn, Brook. Ironies of Oneness and Difference — Coherence in Early Chinese Thought: Prolegomena to the Study of Li 理. Albany: SUNY Press, 2012.

Journal Articles and Book Chapters
Chapter One

Bednar, James A. "Building a mechanistic model of the development and function of the primary visual cortex." *Journal of Physiology - Paris* 106 (2012):194-211.

Block, Ned. "If perception is probabilistic, why does it not seem probabilistic?" *Philosophical Transactions of the Royal Society* B (2018). http://dx.doi.org/10.1098/rstb.2017.0341 .

Corcoran, Andrew W. and Jakob Hohwy. "Allostasis, interoception, and the free energy principle: Feeling our way forward." In *The interoceptive mind: From homeostasis to awareness*. Edited by Manos Tsakiris and Helena De Preester. Oxford University Press, 2018. https://global.oup.com/academic/product/the-interoceptive-mind-9780198811930?q=interoception&lang=en&cc=gb .

de Lange, Floris P., Micha Heilbron, and Peter Kok. "How Do Expectations Shape Perception?" *Trends in Cognitive Science* 22.9 (September 2018):764-79.

Friston, Karl. "Does predictive coding have a future?" *Nature Neuroscience* 21 (August 2018):1019-26.

Hahn, Gerald, Adrian Ponce-Alvarez, Gustavo Deco, Ad Aertsen and Arvind Kumar. "Portraits of communication in neuronal networks." *Nature Reviews Neuroscience* 20 (February 2019):117-27.

Hohwy, Jakob. "New directions in predictive processing." *Mind and Language* 35 (2020):29-23.

———. "New directions in predictive processing." *Mind and Language* 35 (2020):209-23.

Pakkenberg, Bente, Dorte Pelvig, Lisbeth Marner, Mads J. Bundgaard, Hans Jørgen G. Gundersen, Jens R. Nyengaard, and Lisbeth Regeur. "Aging and the Human Neocortex." *Experimental Gerontology* 38 (2003):95-99.

Philips, Ryan T. and V. Srinivasa Chakravarthy. "A Global Orientation Map in the Primary Visual Cortex (V1): Could a Self-Organizing Model Reveal Its Hidden Bias?" *Frontiers in Neural Circuits* 10.109 (January 2017). http://Frontiers in Neural Circuits | www.frontiersin.org, doi: 10.3389/fncir.2016.00109.

Richert, Micah, Dimitry Fisher, Filip Piekniewski, Eugene M. Izhikevich, and Todd L. Hylton. "Fundamental principles of cortical computation: unsupervised learning with prediction, compression and feedback." arXiv:1608.06277 [cs.CV] (August 2016).

Rumelhart, D. E. and D. Zipser. "Feature Discovery by Competitive Learning." In *Parallel Distributed Processing: Explorations in the Microstructure of Cognition, Volume 1: Foundations*. Edited by David E. Rumelhart and James L. McClelland, and PDP Research Group, 170-177. Cambridge: MIT Press, 1986.

Steven J. Olson and Stephen Grossberg, "A neural network model for the development of simple and complex receptive fields within cortical maps of orientation and ocular dominance," *Neural Networks* 11 (1998):189-208.

Yuste, Rafael. "From the neuron doctrine to neural networks." *Nature Reviews: Neuroscience* 16 (August 2015):487-97.

Chapter Two

Bassett, Danielle S. and Olaf Sporns. "Network neuroscience." *Nature Neuroscience* 20 (March 2017):353-64.

Binder, Jeffrey R., Lisa L. Conant, Colin J. Humphries, Leonardo Fernandino, Stephen B. Simons, Mario Aguilar and Rutvik H. Desai. "Toward a brain-based componential semantic representation" *Cognitive Neuropsychology*, 33.3–4 (2016):130–174.

Buckner, Randy L. and Lauren M. DiNicola. "The brain's default network: updated anatomy, physiology and evolving insights." *Nature Reviews Neuroscience* 20 (October 2019):593-608.

Fair, Damien A., Alexander L. Cohen, Jonathan D. Power, Nico U. F. Dosenbach, Jessica A. Church, Francis M. Miezin, Bradley L. Schlaggar, Steven E. Petersen. "Functional brain networks develop from a 'local to distributed' organization." *PLoS Computational Biology* 5.5 (2009): e1000381. https://doi.org/10.1371/journal.pcbi.1000381.

Malvaez, Melissa, Christine Shieh, Michael D. Murphy, Venuz Y. Greenfield and Kate M. Wassum. "Distinct cortical–amygdala projections drive reward value encoding and retrieval." *Nature Reviews Neuroscience* 22 (May 2019):762-69.

Posner, Michael I., Mary K. Rothbart, Brad E. Sheese, and Pascale Voelker. "Control Networks and Neuromodulators of Early Development" *Developmental Psychology* 48.3 (May 2012):827-35.

Pulvermüller, Friedemann. "Neural reuse of action perception circuits for language, concepts and communication." *Progress in Neurobiology* 160 (2018):1-44.

Sporns, Olaf. "Network attributes for segregation and integration in the human brain." *Current Opinion in Neurobiology* 23 (2013):162-71.

Stiles, Joan and Terry L. Jernigan. "The Basics of Brain Development." *Neuropsychology Review* 20 (2010):327-48.

Veale, Richard, Ziad M. Hafed and Masatoshi Yoshida. "How is visual salience computed in the brain? Insights from behaviour, neurobiology and modelling." *Philosophical Transactions of the Royal Society B* (2017). http://dx.doi.org/10.1098/rstb.2016.0113.

Chapter Three

Amso, Dima and Gaia Scerif. "The attentive brain: insights from developmental cognitive neuroscience." *Nature Reviews Neuroscience* 15 (October 2015):606-19.

Baden, Tom, Thomas Euler and Philipp Berens. "Understanding the retinal basis of vision across species." *Nature Reviews Neuroscience* 21 (January 2020):5-20.

Bassett, Danielle S. and Olaf Sporns. "Network neuroscience." *Nature Neuroscience* 20 (March 2017):353-64.

Bassett, Danielle S., Perry Zurn and Joshua I. Gold. "On the nature and use of models in network neuroscience." *Nature Reviews Neuroscience* 19 (September 2018):566-78.

Bickle, John, Marica Bernstein, Matt Heatley, Cindy Worley and Samantha Stiehl. "A Functional Hypothesis for LGN-V1-TRN Connectivities Suggested by Computer Simulation." *Journal of Computational Neuroscience*, 6 (1999):251-261.

Briggs, Farran. "Role of Feedback Connections in Central Visual Processing." *Annual Review of Vision Science* 6.18 (2020):1-22.

Carlo Sestieri, Maurizio Corbetta, Sara Spadone, Gian Luca Romani, and Gordon L. Shulman. "Domain-general signals in the cingulo-opercular network for visuospatial attention and episodic memory." *Journal of Cognitive Neuroscience* 26.3 (March 2014):551-68.

Connor, Charles E and James J Knierim. "Integration of objects and space in perception and memory." *Nature Neuroscience* 20.11 (November 2017):1493-503.

Conway, Bevil R. "The Organization and Operation of Inferior Temporal Cortex." *Annual Review of Vision Science* 4 (2018):381-402.

de Lange, Floris P., Micha Heilbron, and Peter Kok. "How Do Expectations Shape Perception?" *Trends in Cognitive Science* 22.9 (September 2018):764-79.

Dixon, Matthew L., Alejandro De La Vega, Caitlin Mills, Jessica Andrews-Hanna, R. Nathan Spreng, Michael W. Cole, and Kalina Christoff. "Heterogeneity within the frontoparietal control network and its relationship to the default and dorsal attention networks." *PNAS* DOI 10.1073/pnas.1715766115 (January 2018):E1598-1607.

Farrant, Kristafor and Lucina Q. Uddin. "Asymmetric development of dorsal and ventral attention networks in the human brain." *Developmental Cognitive Neuroscience* 12 (2015):165-74.

Gardner, Justin L. "Optimality and heuristics in perceptual neuroscience." *Nature Neuroscience* 22 (April 2019):514-23.

Gilbert, Charles D. and Wu Li. "Top-down Influences on Visual Processing." *Nature Reviews Neuroscience* 14 (May 2013):350-63.

Goodale, Melvyn A. and David Milner. "Separate visual pathways for perception and action." *Trends in Neuroscience* 15.1 (1992): 20–5.

Gordon, Evan M., Charles J. Lynch, Caterina Gratton, Timothy O. Laumann, Adrian W. Gilmore, Deanna J. Greene, Mario Ortega, Annie L. Nguyen, Bradley L. Schlaggar, Steven E. Petersen, Nico U. F. Dosenbach, Steven M. Nelson. "Three Distinct Sets of Connector Hubs Integrate Human Brain Function." *Cell Reports* 24.7 (August 2018):1687-95.

Gratton, Caterina, Haoxin Sun, and Steven E. Petersen. "Control Networks and Hubs." *Psychophysiology* 55.3 (March 2018), DOI 10-1111/psyp.13032.

Haak, Koen V. and Christian F. Beckmann. "Objective analysis of the topological organization of the human cortical visual connectome suggests three visual pathways." *Cortex* 98 (2018):73-83.

Halassa, Michael M. and Sabine Kastner. "Thalamic functions in distributed cognitive control." *Nature Neuroscience* 20 (December 2017):1669-79.

Hasse, J. Michael and Farran Briggs. "Corticogeniculate feedback sharpens the temporal precision and spatial resolution of visual signals in the ferret." *PNAS* July 11, 2017. www.pnas.org/cgi/doi/10.1073/pnas.1704524114.

Horga, Guillermo and Anissa Abi-Dargham. "An integrative framework for perceptual disturbances in psychosis." *Nature Reviews Neuroscience* 20 (December 2019):763-78.

Kourtzi, Zoe, Mark Augath, Nikos K. Logothetis, J. Anthony Movshon, and Lynne Kiorpes. "Development of visually evoked cortical activity in infant macaque monkeys studied longitudinally with fMRI." *Magnetic Resonance Imaging* 24 (2006):359–366.

Kravitz, Dwight J., Kadharbatcha S. Saleem, Chris I. Baker, Leslie G. Ungerleider, and Mortimer Mishkin. "The ventral visual pathway: an expanded neural framework for the processing of object quality." *Trends in Cognitive Science* 17.1 (January 2013):26-49.

Liang Hualou, Xiajing Gong, Minggui Chen, Yin Yan, Wu Li, and Charles D. Gilbert. "Interactions between feedback and lateral connections in the primary visual cortex." *PNAS* 114.32 (August 8, 2017):8637-42.

Pearson, Joel. "The human imagination: the cognitive neuroscience of visual mental imagery." *Nature Reviews Neuroscience* 20 (October 2019):624-34.

Posner, Michael I. and Steven E. Petersen. "The attention system of the human brain." *Annual Review of Neuroscience* 13 (1990):25–42.

Ptak, Radek. "The Frontoparietal Attention Network of the Human Brain: Action, Saliency, and a Priority Map of the Environment." *The Neuroscientist* 18.5 (October 2012):502-15.

Rosen, Maya L., Margaret A. Sheridan, Kelly A. Sambrook, Matthew R. Peverill, Andrew N. Meltzoff, and Katie A. McLaughlin. "The Role of Visual Association Cortex in Associative Memory Formation across Development." *Journal of Cognitive Neuroscience* 30.3 (March 2018):365-80.

Veale, Richard, Ziad M. Hafed and Masatoshi Yoshida. "How is visual salience computed in the brain? Insights from behaviour, neurobiology and modelling." *Philosophical Transactions of the Royal Society B* (2017)1-14.

Webb, Taylor W., Kajsa M. Igelström, Aaron Schurger, and Michael S. A. Graziano. "Cortical networks involved in visual awareness independent of visual attention." *PNAS* 113.48 (November 29, 2016):13923-28.

Chapter Four

Adolphs, Ralph. "How should neuroscience study emotions? by distinguishing emotion states, concepts, and experiences." *Social Cognitive and Affective Neuroscience* (2017): 24-31

Barrett, Lisa Feldman. "The theory of constructed emotion: an active inference account of interoception and categorization." *Social Cognitive and Affective Neuroscience* 2017:1-23.

___ and Ajay B. Satpute. "Historical pitfalls and new directions in the neuroscience of emotion." *Neuroscience Letters* (2017):1-10.

___, Christine D. Wilson-Mendenhall, and Lawrence W. Barsalou. "The Conceptual Act Theory: A Roadmap." In *The Psychological Construction of Emotion*, 83-110.

Carter, Sue and Stephen W. Porges. "Neurobiology and the Evolution of Mammalian Social Behavior." In *Evolution, Early Experience and Human Development*, 32-51.

Cunningham, William A., Kristen Dunfield, and Paul E. Stillman. "Affect Dynamics: Iterative Reprocessing in the Production of Emotional Responses." In *The Psychological Construction of Emotion*, edited by Lisa Feldman Barrett and James A. Russell, 168-82. New York: Guilford Press, 2015.

Davis, Kenneth L. and Christian Montag. "Selected Principles of Pankseppian Affective Neuroscience." *Frontiers in Neuroscience* 12 (January 2019) doi: 10.3389/fnins.2018.01025.

Ekman, Paul and Daniel Cordaro. "What is Meant by Calling Emotions Basic." *Emotion Review* 7.4 (October 2015):364-70.

Feldman, Ruth. "The Neurobiology of Human Attachments." *Trends in Cognitive Sciences* 21.2 (February 2017):80-99.

___, Katharina Braun and Frances A. Champagne. "The neural mechanisms and consequences of paternal caregiving." *Nature Reviews: Neuroscience* 20 (April 2019):205-24.

Harris, Haley N. and Yuan B. Peng. "Evidence and explanation for the involvement of the nucleus accumbens in pain processing." *Neural Regeneration Research* 15.4 (October 2019):597-605.

Hoemann, Katie, Maria Gendron, and Lisa Feldman Barrett. "Mixed emotions in the predictive brain." *Current Opinion in Behavioral Science* 15 (June 2017):51-57.

James, William. "What is an Emotion?" *Mind* 9.34 (April 1884):188-205.

Kleckner, Ian R., and Karen S. Quigley. "An Approach to Mapping the Neurophysiological State of the Body to Affective Experience." In *The Psychological Construction of Emotion*, 265-301.

MacCormack, Jennifer K. and Kristen A. Lindquist. "Bodily Contributions to Emotion: Schachter's Legacy for a Psychological Constructionist View on Emotion." *Emotion Review* 9.1 (January 2017):36-45.

Nelson, Eric E. "The Neurobiological Basis of Empathy and Its Development in the Context of Our Evolutionary Heritage." in *Evolution, Early Experience and Human Development*, 179-98.

Panksepp, Jaak. "Cross-Species Affective Neuroscience Decoding of the Primal Affective Experiences of Humans and Related Animals." *PLoS One* 6.9 (September 2011):1-15.

___. "How Primary-Process Emotional Systems Guide Child Development: Ancestral Regulators of Human Happiness, Thriving, and Suffering." In *Evolution, Early Experience and Human Development*, 74-94.

___. "The basic emotional circuits of mammalian brains: Do animals have affective lives?" *Neuroscience and Biobehavioral Review* 35 (2011):1791-1804.

___, Mark Solm, Richard D. Lane, and Ryan Smith. "Reconciling cognitive and affective neuroscience perspectives on the brain basis of emotional experience." *Neuroscience and Biobehavioral Review* 76 (2017):187-215.

Paul, Elizabeth S., Shlomi Sher, Marco Tamietto, Piotr Winkielman, and Michael T. Mendl. "Towards a comparative science of emotion: Affect and consciousness in humans and animals." *Neuroscience and Biobehavioral Reviews* 108 (2020):749-770.

Pessoa, Luiz. "Emotion and the Interactive Brain: Insights from Comparative Neuroanatomy and Complex Systems." *Emotion Review* 10.3 (July 2018):204–216.

___. "Understanding emotion with brain networks." *Current Opinion in Behavioral Science* 19 (2018 February):19–25.

Sander, David, Didier Grandjean and Klaus R. Scherer. "An Appraisal-Driven Componential Approach to the Emotional Brain." *Emotion Review* 10.3 (July 2018):219-31.

Seth, Anil K. "Consciousness: The last 50 years (and the next)." *Brain and Neuroscience Advances* 2 (2018):1-6.

___. "Interoceptive inference, emotion, and the embodied self." *Trends in Cognitive Sciences*, Vol. 17, No. 11 (November 2013):565-73.

___ and Manos Tsakiris. "Being a Beast Machine: The Somatic Basis of Selfhood." *Trends in Cognitive Sciences*, 22.11 (November 2018):969-81.

Siegel, Daniel J. "The Integrative Meaning of Emotion." In *Evolution, Early Experience and Human Development: From Research to Practice and Policy*, edited by Darcia Narvaez, Jaak Panksepp, Allan N. Schore, and Tracy R. Gleason, 95-98. Oxford: Oxford University Press, 2013.

Smith, Ryan and Richard D. Lane. "The neural basis of one's own conscious and unconscious emotional states." *Neuroscience and Biobehavioral Review* 57 (2015):1-29.

Todd, Rebecca M., Vladimir Miskovic, Junichi Chikazoe, and Adam K. Anderson. "Emotional Objectivity: Neural Representations of Emotions and Their Interaction with Cognition." *Annual Review of Psychology* 71 (2020):25-48.

Chapter Five

Amso, Dima and Gaia Scerif. "The attentive brain: insights from developmental cognitive neuroscience." *Nature Reviews Neuroscience* 15 (October 2015):606-19.

Buckner, Randy L. and Lauren M. DiNicola. "The brain's default network: updated anatomy, physiology and evolving insights." *Nature Reviews Neuroscience* 20 (October 2019):593-608.

Bulgarelli, Chiara, Anna Blasi, Carina C.J.M. de Klerk, John E. Richards, Antonia Hamilton, and Victoria Southgate. "Fronto-temporoparietal connectivity and self-awareness in 18-month-olds: A resting state fNIRS study." *Developmental Cognitive Neuroscience* 38 (2019):1-12.

Cassidy, Jude. "The Nature of the Child's Ties." In *Handbook of Attachment*, 3-24.

De Asis-Cruz, Josepheen, Marine Bouyssi-Kobar, Iordanis Evangelou, Gilbert Vezina, and Catherine Limperopoulos. "Functional properties of resting state networks in healthy full-term newborns." *Nature: Scientific Reports* (December 2015) DOI: 10.1038/srep17755:1-15.

Dean, Douglas C. III, Jonathan O'Muircheartaigh, Holly Dirks, Nicole Waskiewicz, Lindsay Walker, Ellen Doernberg, Irene Piryatinsky, and Sean C. L. Deoni. "Characterizing longitudinal white matter development during early childhood." *Brain Structure and Function* 220 (2015):1921-33.

Delafield-Butt. Jonathan T. and Nivedita Gangopadhyay. "Sensorimotor intentionality: The origins of intentionality in prospective agent action." *Developmental Review* 33 (2013) 399–425.

Ellis, Cameron T., Lena J. Skalaban, Tristan S. Yates, Vikranth R. Bejjanki, Natalia I. Córdova, and Nicholas B. Turk-Browne. "Evidence of hippocampal learning in human infants." *Current Biology* 31 (Aug. 2021):1-7.

Feldman, Ruth. "The Neurobiology of Human Attachments." *Trends in Cognitive Sciences* 21.2 (February 2017):80-99.

Gilmore, John H., Rebecca C. Knickmeyer and Wei Gao. "Imaging structural and functional brain development in early childhood." *Nature Reviews Neuroscience* 19 (March 2018):123-37.

Henderson, Heather A. and Peter C. Mundy. "The Integration of Self and Other in the Development of Self-Regulation: Typical and Atypical Processes." In *Handbook of Self-Regulatory Processes in Development New Directions and International Perspectives: The Integration of Self and Other in the Development of Self-Regulation*, edited by Karen Caplovitz Barrett, Nathan A. Fox, George A. Morgan, Deborah J. Fidler, Lisa A. Daunhauer. Published online by Routledge on: 17 Dec 2012.

Hodel, Amanda S. "Rapid infant prefrontal cortex development and sensitivity to early environmental experience." *Developmental Review* 48 (2018):113–144.

Kelly, Michael P., Natasha M. Kriznik, Ann Louise Kinmonth and Paul C. Fletcher. "The brain, self and society: a social-neuroscience model of predictive processing." *Social Neuroscience*, 14.3 (2019):266-276, DOI: 10.1080/ 17470919.2018.1471003.

Keunen, Kristin, Serena J. Counsell, and Manon J.N.L. Benders. "The emergence of functional architecture during early brain development." *NeuroImage* 160 (2017):2-14.

Knudsen, Eric I. "Sensitive Periods in the Development of the Brain and Behavior." *Journal of Cognitive Neuroscience* 16.8 (October 2004):1412-25.

Marvin, Robert S., Preston A. Britner, and Beth S. Russell. "Normative Development – The Ontogeny of Attachment in Childhood." In *Handbook of Attachment: Theory, Research, and Clinical Applications*, edited by Jude Cassidy and Phillip R. Sharver, 3rd edition, 273-90. New York: The Guilford Press, 2016.

Polan, H. Jonathan and Myron A. Hofer. "Psychobiological Origins of Infant Attachment and Its Role in Development." In *Handbook of Attachment*, 117-32.

Posner, Michael I., Mary K. Rothbart, Brad E. Sheese, and Pascale Voelker. "Control Networks and Neuromodulators of Early Development." *Developmental Psychology* 48.3 (May 2012):827-35.

Raz, Gal and Rebecca Saxe. "Learning in Infancy Is Active, Endogenously Motivated, and Depends on the Prefrontal Cortices." *Annual Review of Developmental Psychology* 2020.2:247-68.

Rothbart, Mary K., Brad E. Sheese, M. Rosario Rueda, and Michael I. Posner. "Developing Mechanisms of Self-Regulation in Early Life." *Emotion Review* 3.2 (April 2011):207–213. doi:10.1177/1754073910387943.

Sherman, Laura J., Katherine Rice, and Jude Cassidy. "Infant capacities related to building internal working models of attachment figures: A theoretical and empirical review." *Developmental Review* 37 (2015):109-41.

Sperry, Roger W. "Neurology and the mind–brain problem." *American Scientist*, 40 (1952):291–312.

Sroufe, L. Alan. "Attachment and development: A prospective longitudinal study from birth to adulthood." *Attachment and Human Development* 7 (2005):34-80.

Stiles, Joan and Terry L. Jernigan. "The Basics of Brain Development." *Neuropsychology Review* 20 (2010):327–348.

Stiles, Joan, Timothy T. Brown, Frank Haist, and Terry L. Jernigan. "Brain and Cognitive Development." In *Handbook of Child Psychology and Developmental Science, 7th Edition*, edited by Richard M. Lerner, 1-54. John Wiley & Sons, Inc., 2015.

Tau, Gregory Z. and Bradley S Peterson. "Normal Development of Brain Circuits." *Neuropsychopharmacology Reviews* (2010):147–168.

Yeshurun, Yaara, Mai Nguyen and Uri Hassan. "The default mode network: where the idiosyncratic self meets the shared social world." *Nature Reviews Neuroscience* 22 (March 2021):181-92.

Chapter Six

Adamantidis, Antoine R., Carolina Gutierrez Herrera and Thomas C. Gent. "Oscillating circuitries in the sleeping brain." *Nature Reviews Neuroscience* 20 (December 2019):746-62.

Albertazzi, Liliana. "Naturalizing Phenomenology: A *Must Have?*" *Frontiers in Psychology*, published 22 October 2018, doi: 10.3389/fpsyg.2018.01933.

Barker, Gareth R. I., Paul J Banks, Hannah Scott, G Scott Ralph, Kyriacos A Mitrophanous, Liang-Fong Wong, Zafar I Bashir, James B Uney and E Clea Warburton. "Separate elements of episodic memory subserved by distinct hippocampal–prefrontal connections." *Nature Neuroscience* 20.2 (February 2017):242-50.

Barsalou, Lawrence W. "Challenges and Opportunities for Grounding Cognition." *Journal of Cognition*, 3.1 (2020):1–24.

Bauer, Patricia J. "Development of episodic and autobiographical memory: The importance of remembering forgetting." *Developmental Review* 38 (2015): 146–166.

Borbély, Alexander A., Serge Daan, Anna Wirz-Justice and Tom Deboer. "The two-process model of sleep regulation: a reappraisal." *Journal of Sleep Research* 25 (2016):131-43.

Carota, Francesca, Nikolaus Kriegeskorte, Hamed Nili and Friedemann Pulvermüller. "Representational Similarity Mapping of Distributional Semantics in Left Inferior Frontal, Middle Temporal, and Motor Cortex." *Cerebral Cortex* 27 (January 2017):294-309.

Chen, Janice, Yuan Chang Leong, Christopher J Honey, Chung H Yong, Kenneth A Norman and Uri Hasson. "Shared memories reveal shared structure in neural activity across individuals." *Nature Neuroscience* 20.1 (January 2017): 115-24.

Deboer, Tom. "Sleep homeostasis and the circadian clock: Do the circadian pacemaker and the sleep homeostat influence each other's functioning?" *Neurobiology of Sleep and Circadian Rhythms* 5 (2018):68–77.

Ellis, Cameron T., Lena J. Skalaban, Tristan S. Yates, Vikranth R. Bejjanki, Natalia I. Córdova, Nicholas B. Turk-Browne. "Evidence of hippocampal learning in human infants." *Current Biology* 31 (Aug. 2021):1-7.

Favila, Serra E., Hongmi Lee, and Brice A. Kuhl. "Transforming the Concept of Memory Reactivation." *Trends in Neuroscience* 43.12 (December 2020):939-50.

Friederici, Angel D. "The Neuroanatomical Pathway Model of Language: Syntactic and Semantic Networks." In *Neurobiology of Language*, 349-56. DOI: http://dx.doi.org/10.1016/B978-0-12-407794-2.00029-8.

Fuster, Joaquin. "Upper Processing Stages of the Perception-Action Cycle." *Trends in Cognitive Science* 8.4 (April 2004):143-45.

Garagnani, Max, Evgeniya Kirilina, and Friedemann Pulvermüller. "Semantic Grounding of Novel Spoken Words in the Primary Visual Cortex." *Frontiers of Human Neuroscience: Speech and Language*, February 24, 2021, doi: 10.3389/fnhum.2021.581847.

Gava, Giuseppe P., Stephen B. McHugh, Laura Lefèvre, Vítor Lopes-dos-Santos, Stéphanie Trouche, Mohamady El-Gaby, Simon R. Schultz and David Dupret. "Integrating new memories into the hippocampal network activity space." *Nature Neuroscience* 24.3 (March 2021):326-30.

Gómez, Rebecca L. and Jamie O. Edgin. "The extended trajectory of hippocampal development: Implications for early memory development and disorder." *Developmental Cognitive Neuroscience* 18 (2016):57–69.

Hainmueller, Thomas and Marlene Bartos. "Dentate gyrus circuits for encoding, retrieval and discrimination of episodic memories." *Nature Reviews Neuroscience* 21.3 (March 2020):153-68.

Harnad, Stevan. "The Symbol Grounding Problem." *Physica D: Nonlinear Phenomena* 42.1-3 (June 1990):335-46.

Hickok, Gregory and David Poeppel. "Neural Basis of Speech Perception." In *Neurobiology of Language*, 299-310. DOI: http://dx.doi.org/10.1016/B978-0-12-407794-2.00025-0.

Joo, Hannah R. and Loren M. Frank. "The hippocampal sharp wave–ripple in memory retrieval for immediate use and consolidation." *Nature Reviews Neuroscience* 19 (December 2018):744-57.

Klinzing, Jens G., Niels Niethard and Jan Born. "Mechanisms of systems memory consolidation during sleep." *Nature Neuroscience* 22 (October 2019):1598-1610.

Le Duigou, Caroline, Jean Simonnet, Maria T. Teleńczuk, Desdemona Fricker and Richard Miles. "Recurrent synapses and circuits in the CA3 region of the hippocampus: an associative network." *Frontiers in Cellular Neuroscience* 7 (January 2014), doi: 10.3389/fncel.2013.00262.

Li, Wei, Lei Ma, Guang Yang and Wen-Biao Gan. "REM sleep selectively prunes and maintains new synapses in development and learning." *Nature Neuroscience* 20.3 (March 2017):427-37.

Lopopolo, Alessandro, Antal van den Bosch, Karl-Magnus Petersson, and Roel M. Willems. "Distinguishing Syntactic Operations in the Brain: Dependency and Phrase-Structure Parsing." *Neurobiology of Language* 2.1 (January 2021):152-75.

Mascetti, Gian Gastone. "Unihemispheric sleep and asymmetrical sleep: behavioral, neurophysiological, and functional perspectives." *Nature and Science of Sleep* 8 (2016):221-38.

Masukar, Arjun V. "Towards a Circuit-Level Understanding of Hippocampal CA1 Dysfunction in Alzheimer's Disease Across Anatomical Axes." *Journal of Alzheimer's Disease and Parkinsonism* 8.1 (2018), DOI: 10.4172/2161-0460.1000412.

McClelland, James L., Bruce L. McNaughton, and Randall C. O'Reilly. "Why There Are Complementary Learning Systems in the Hippocampus and Neocortex: Insights from the Successes and Failures of Connectionist Models of Learning and Memory." *Psychological Review* 102.3 (1995): 419-57.

McCloskey, Michael and Neal J. Cohen. "Catastrophic Interference in Connectionist Networks: The Sequential Learning Problem." *Psychology of Learning and Motivation* 24 (1989):109-65.

Mullally, Sinéad L. "Commentary: Elucidating the neural correlates of early childhood memory." *International Journal of Behavioral Development* 39.4 (2015):306-07.

___ and Eleanor A. Maguire. "Learning to remember: The early ontogeny of episodic memory." *Developmental Cognitive Neuroscience* 9 (2014):12-29.

Niethard, Niels and Jan Born. "Back to baseline: sleep recalibrates synapses." *Nature Neuroscience* 22 (February 2019):149-53.

Nilssen, Eirik S., Thanh P. Doan, Maximiliano J. Nigro, Shinya Ohara and Menno P. Witter. "Neurons and networks in the entorhinal cortex: A reappraisal of the lateral and medial entorhinal subdivisions mediating parallel cortical pathways." *Hippocampus.* 29 (2019):1238–1254.

O'Mara, Shane. "The subiculum: what it does, what it might do, and what neuroanatomy has yet to tell us." *Journal of Anatomy* 207 (2005): 271-82.

Patterson, Karalyn and Matthew A. Lambon Ralph. "The Hub-and-Spoke Hypothesis of Semantic Memory." In *Neurobiology of Language*, 765-776.

Rabola, Nelson, Mario Carta, and Christophe Mulle. "Operation and plasticity of hippocampal CA3 circuits: implications for memory encoding." *Nature Reviews Neuroscience* 18.4 (April 2017): 209-21.

Ramsaran, Adam I., Margaret L. Schlichting, and Paul W. Frankland. "The ontogeny of memory persistence and specificity." *Developmental Cognitive Neuroscience* 36 (2019):1-15.

Raslau, F.D., I.T. Mark, A.P. Klein, J.L. Ulmer, V. Mathews and L.P. Mark. "Memory Part 2: The Role of the Medial Temporal Lobe." In *American Journal of Neuroradiology* 2015, 36.5 (May):846-849; DOI: https://doi.org/10.3174/ajnr.A4169.

Raz, Gal and Rebecca Saxe. "Learning in Infancy Is Active, Endogenously Motivated, and Depends on the Prefrontal Cortices." *Annual Review of Developmental Psychology* 2020.2:247-68.

Rissman, Lilia and Asifa Majid. "Thematic roles: Core knowledge or linguistic construct?" *Psychonomic Bulletin & Review* 26 (2019):1850–69.

Saper, Clifford B. and Patrick M. Fuller. "Wake-sleep circuitry: an overview." *Current Opinion in Neurobiology* 44 (2017):186-192.

Sassenhagen, Jona and Christian J. Fiebach. "Traces of Meaning Itself: Encoding Distributional Word Vectors in Brain Activity." *Neurobiology of Language* 1.1 (January 2020):54-76.

Schapiro, Anna C., Nicholas B. Turk-Browne, Matthew M. Botvinick and Kenneth A. Norman. "Complementary learning systems within the hippocampus: a neural network modelling approach to reconciling episodic memory with statistical learning." *Philosophical Transactions B of the Royal Society* (2016) http://dx.doi.org/10.1098/rstb.2016.0049).

Shauna M. Stark and Craig E. L. Stark, "Introduction to Memory." In *Neurobiology of Language*, edited by Greg Hickok and Steve Small, 841-54. Academic Press, 2015. DOI: http://dx.doi.org/10.1016/B978-0-12-407794-2.00067-5

Squire, Larry and Pablo Alvarez. "Retrograde amnesia and memory consolidation: a neurobiological perspective." *Current Opinion in Neurobiology* 5 (1995):169-77.

Staples, Ryan and William W. Graves. "Neural Components of Reading Revealed by Distributed and Symbolic Computational Models." *Neurobiology of Language* 1.4 (September 2020):381-401.

Sun, Yanjun, Suoqin Jin, Xiaoxiao Lin, Lujia Chen, Xin Qiao, Li Jiang, Pengcheng Zhou, Kevin G. Johnston, Peyman Golshani, Qing Nie, Todd C. Holmes, Douglas A. Nitz and Xiangmin Xu. "CA1-projecting subiculum neurons facilitate object–place learning." *Nature Neuroscience* 22.11 (November 2019): 1857-70.

Tamaki, Masako, Zhiyan Wang, Tyler Barnes-Diana, DeeAnn Guo, Aaron V. Berard, Edward Walsh, Takeo Watanabe and Yuka Sasaki. "Complementary contributions of non-REM and REM sleep to visual learning." *Nature Neuroscience* 23.9 (September 2020):1150-56.

Tang, Evelyn, Marcelo G. Mattar, Chad Giusti, David M. Lydon-Staley, Sharon L. Thompson-Schill and Danielle S. Bassett. "Effective learning is accompanied by high-dimensional and efficient representations of neural activity." *Nature Neuroscience* 22 (July 2019):1000-09.

Tompary, Alexa and Lila Davachi. "Consolidation Promotes the Emergence of Representational Overlap in the Hippocampus and Medial Prefrontal Cortex." *Neuron* 96 (September 27, 2017):228-41.

Tononi, Giulio and Chiara Cirelli. "Sleep and synaptic down-selection." *European Journal of Neuroscience* 2020 (51):413-21.

Tononi, Giulio and Chiara Cirelli. "Sleep and the Price of Plasticity: From Synaptic and Cellular Homeostasis to Memory Consolidation and Integration." *Neuron* 81 (January 8, 2014):12-34.

Tubbs, Andrew S., Hannah K. Dolish, Fabian Fernandez, and Michaél A. Grandner. "The basics of sleep physiology and behavior." In *Sleep and Health*, edited by Michael Grandner, 3-10. Cambridge: Academic Press, 2019. https://doi.org/10.1016/B978-0-12-815373-4.00001-0.

Tulving, Endel. "Episodic and semantic memory." In *Organization of Memory edited by Endel Tulving and Wayne Donaldson*, 381-402. New York: Academic Press, 1972.

Tuncdemir, Sebnum Nur, Clay Orion Lacefield, and Rene Hen. "Contributions of adult neurogenesis to dentate gyrus network activity and computations." *Behavioral Brain Research* 374 (2019) 112112, 12 pages. https://doi.org/10.1016/j.bbr.2019.112112.

Wimmer, G. Elliott, Yunzhe Liu, Neža Vehar, Timothy E. J. Behrens and Raymond J. Dolan. "Episodic memory retrieval success is associated with rapid replay of episode content." *Nature Neuroscience* 23.8 (August 2020):1025-1033.

Xiao, Xiaoqian, Qi Dong, Jiahong Gao, Weiwei Men, Russell A. Poldrack, and Gui Xue. "Transformed Neural Pattern Reinstatement during Episodic Memory Retrieval." *Journal of Neuroscience* 37.11(March 15, 2017):2986-98.

Chapter Seven

Aragona, Massimiliano. "Phenomenology, Naturalism, and the Neuroscience." In *The Oxford Handbook of Phenomenological Psychopathology*. edited by Giovanni Stanghelline et al., 273-83. Oxford: Oxford University Press, 2019.

Berthrong, John H. "Religion in the *Xunzi*: What Does *Tian* 天 Have to Do with It?" In *Dao Companion to the Philosophy of Xunzi*, edited by Eric L. Hutton, 323-51. Dordrecht: Springer, 2016.

Carroll, Thomas D. "Wittgenstein and the *Xunzi* on the Clarification of Language." *Dao* 17 (2018):527-45.

Fraser, Chris. "Language and Logic in the *Xunzi*." In *Dao Companion to the Philosophy of Xunzi*, 291-321.

Gallagher, Shaun. "On the possibility of naturalizing phenomenology." In *The Oxford Handbook of Contemporary Phenomenology*, edited by Dan Zahavi. Oxford: Oxford University Press, 2012. DOI: 10.1093/oxfordhb/9780199594900.013.0005.

Hutton, Eric L. "Xunzi on Moral Psychology." In *Dao Companion to the Philosophy of Xunzi*, 201-27.

Nylan, Michael. "Xunzi: An Early Reception History, Han Through Tang." In *Dao Companion to the Philosophy of Xunzi*, 395-433.

Perkins, Franklin. "Music and Affect: The Influence of the *Xing Zi Ming Chu* on the *Xunzi* and *Yueji*." *Dao* 16 (2017):325-40.

Robins, Dan. "The Warring States Concept of *Xing*." *Dao* 10 (2011):31-51.

Schulzki, Irina. "Love at Loss: Jen-Luc Marion's Concept of Erotic Reduction and Paul Thomas Anderson's *Magnolia*." In *Einheitsdenken nach der Postmoderne. Figuren von Ganzheit, Präsenz und Transzendenz nach der Postmoderne*, edited by Irina Hron, 145-71. Nordhausen: Verlag Traugott Bautz, 2015.

Index

Ainsworth, Mary, 163-66
allostasis, 43-44, 106, 123
altricial animals, 137, 161
amygdala, 53, 55, 118
attachment, 157, 161-66
attentional networks, 149-51
attractor networks, 35-39, 185

Barrett, Lisa Feldman, 102, 106-07, 121ff
basal ganglia, 53f
Bayesian inference, 41ff, 88
Boddice, Rob, 130-33
Bowlby, John, 161-62
brainstem, 50, 102, 110ff, 122, 154-56
Brodmann areas, 59-61
Brodmann, Korbinian (1868-1918), 59
Buddhism, 1-2

Cao Pi (187-226), 235-236
Cao Zhi (192-232), "Seeing off Mr. Ying," 236-37
catastrophic interference, 172-74
caudal, 57-58
cell migration, 137ff
central sulcus, 56-58
cerebellum, 50
cerebral cortex, 58; grey matter, 58; white matter, See White matter

chemistry, 4
Chinese poetic tradition, 229
cingulate cortex, 57-58; anterior cingulate cortex, 118
Clark, Andy, 45-47
coarse coding, 77
combinatorial (sentential) semantics, 210
computer programming analogy, 17
connectome, 66

Damasio, Antonio, 102, 105, 109 fn. 26, 115ff; autobiographical self, 127-28; core self, 116, 126; dispositional memory, 128; proto-self, 115-16, 126
default mode network (DMN), 65, 151-153, 201-02
dentate gyrus, 183
dependency grammar, 213-14
development, apoptosis, 141; dendritic arborization, 142; epigenetic-driven, 138ff; experience-dependent, 132, 141; experience-expectant, 145; order of myelination, 145; sensitive periods, 145-46; synaptogenesis, 141ff
Dilthey, Wilhelm (1833-1911), 133-35

dispositional matrices, 102-03, 106-14
distributional semantics, 208-09
dorsal, 57-58
Du Fu (713-770), "Weary Night," 237-40

Ekland, Paul, 102
embodied (or grounded) semantics, 205-08
embodiment, 12
embryo, 137ff
emotion, 98; basic emotions, 98; constructionist model, 121; history of emotions, 129-35; primary-process emotions, 104ff
emotional communities, 130-131
emotional concepts, 121ff
entorhinal cortex, 182ff
episodic memory, 87, 171ff, 194ff

Falke, Cassandra, 221-228
feedforward networks, 69, 76
feelings, 120; as felt experience, 124, 126-27; *qing* 情, 6 fn. 6, 231f
Feldman, Ruth, 112-114
forebrain, 53
foveation, 72, 171; also see saccade
Friston, Karl, 41, 44
frontal lobe, 56-58
functional magnetic resonance imaging (fMRI), 64, 201

graph theory, 65, 148
Graziano, Michael, 124
"Great Preface" to the *Canon of Poetry*, 234-35

Harnad, Stefan, 205-206

Hebb, Donald O. (1904-1985), 17; Hebb's learning rule, 20-21, 24
hippocampus, 53, 55, 182; CA1, 187-89; CA3, 185-87; subiculum, 188-189
history of emotion, 129-35
homeostasis, 43-44, 106
homunculus, 61-62
Husserl, Edmund, 221ff

inner working model (IWM), 161ff
insula, 57-58, 117-18
intentions, 100, 168-69; sensorimotor intentionality, 154-156, 160; *yi* 意, 6 fn. 6, 232f
interoception, 103, 115ff, 122, 158

joint attention, 164; initiate joint attention" (IJA), 165-166; response to joint attention" (RJA), 164

Kant, Immanuel, 2 fn. 3
Koch, Christof, 124

language processing, 211-16
lateral, 57-58
lateral geniculate nucleus (LGN), 25-26, 30, 76-80
limbic lobe, see cingulate cortex

Marion, Jean-Luc, 221-28
matrix, 23
McClelland, James, 172-74
meaning, 12
medial, 57-58
medial temporal cortical memory system, 86, 179-89
medial temporal lobe, 179

memory, complementary learning systems, 175, 189ff; declarative, 179; explicit, 179; implicit, 179
memory consolidation, 174, 194-201
midbrain, 50-53, 102
multivoxel pattern analysis (MVPA), 64, 201
myelination, 141, 144-46

nature (*xing* 性), 6 fn. 6, 231ff
neocortex, 50, 55
network neuroscience, 64-67, 152
neural network modeling, 19-25; adaptive resonance model, 37-39; back-propagation of error, 28-29, 33; feature detectors, 30; figure-ground separation, 37; hidden units, 26-29; Kohonen SOM network, 31-32; lateral connectivity, 30; recurrent networks, 33-35: self-organizing maps, 29-33; top-down connectivity, 35: topographic map, 25, 33
neural progenitor cells, 137ff
neuromodulators, 52, 102-03, 108ff
neurons, 14-16; depolarization, 14-15; spiking, 14-16, 39-40
neurotransmitters, 15; excitatory, 15; inhibitory, 15, 24

occipital lobe, 56-58; also see visual cortex
orbitofrontal cortex, 57-58
oxytocin, 111-13, 160

Panksepp, Jaak (1943-2017), 98 fn. 5, 99ff, 104ff
parahippocampal cortex, 180, 182

parietal lobe, 56-58, 90ff
penguins, 173
perceptron, 19-20, 33; limitations, 27
periaqueductal grey (PAG), 52, 110
perirhinal cortex, 181
Pessoa, Luiz, 119
phenomenal realm, 2
phenomenology, 221-28
Plamper, Jan, 98 fn. 2, 130
Posner, Michael, 84; and Mary Rothbart, 149
predictive processing, 40-48, 79-80, 88, 121-22, 177ff
prefrontal cortex, 57-58
primary visual cortex (V1), 25-26, 30, 60-62, 80-87; retinotopic map, 93; also see visual system
proprioception, 109, 154

REM sleep, 191-93
resolve (*zhi* 志), 6 fn. 6, 232ff
retina, 70-75; bipolar cells, 74; color perception, 73; fovea, 71, 83; ganglion cells, 74-76; horizontal cells, 74; rods and cones, 70
Rosenwein, Barbara H., 130-131
rostral, 57-58

saccade, 72
salience, 102-03; salience network, 84-85, 147
semantic memory, 103, 171ff, 194ff, 202-04
Seth, Anil K., 125-26
Siddhartha Gautama, 1
sigma notation, 22
sleep cycle, 190-193
Smith, Ryan and Richard D. Lane, 116-18, 125

social self, 147, 156
Sporns, Olaf, 66
"strange situation", 165-66
structuralism, 5
Su Shi (1037-1101), 218
superior colliculus, 51-52
symbol representation, 205-06
synaptic junction, 16, 184

temporal lobe, 56-58; also see medial temporal lobe
thalamus, 53, 140-41, 191ff; also see lateral geniculate nucleus

valence, 102-03, 160, 182
vector, 21-23
ventral, 57-58
ventral tegmental area (VTA), 52, 109-10, 160

visual system: attention, 77, 82; dorsal pathway, 90; feedback connectivity, 76, 86; fusiform face area (FFA), 88; Hebbian learning, 78; inferotemporal cortex (inferior temporal cortex), 93-95; ventral pathway, 90-95
voxel, 64

weighting matrix, 21
white matter, 59, 64, 92, 212ff
words, 205-208

Xun Kuang (ca. 310-235 BCE), 229-33

Zhuang Zhou (ca. 369-286 BCE), 229

www.ingramcontent.com/pod-product-compliance
Lightning Source LLC
Chambersburg PA
CBHW081202240426
43669CB00039B/2766